Guide to
PICMICRO®
Microcontrollers

Guide to
PICMICRO®
Microcontrollers

CARL BERGQUIST

©2001 by Sams Technical Publishing

PROMPT® Publications is an imprint of Sams Technical Publishing, 5436 W. 78th St., Indianapolis, IN 46268.

International Standard Book Number: 0-7906-1217-8
Library of Congress Catalog Card Number: 00-111344

Acquisitions Editor: Alice J. Tripp
Editor: Kim Heusel
Assistant Editor: Will Gurdian
Typesetting: Will Gurdian and Kim Heusel
Cover Design and Graphics Conversion: Christy Pierce
Illustrations and other materials: Courtesy the author and Microchip Technology, Incorporated. Reprinted with permission of the copyright owner, Microchip Technology Incorporated. All rights reserved.

PRINTED IN THE UNITED STATES OF AMERICA

9 8 7 6 5 4 3 2 1

ACKNOWLEDGMENTS

I would like to thank the following for their contribution to this book. Your help was valuable and greatly appreciated.

• Microchip Technology Incorporated. All Microchip Technology materials are reprinted here with permission of the copyright owner, Microchip Technology Incorporated. All rights reserved. No further reprints or reproductions may be made without Microchip Technology's prior written consent. Information contained in this publication regarding device applications and the like is intended as suggestion only and may be superseded by updates. No representation or warranty is given, and no liability is assumed by Microchip Technology Incorporated with respect to the accuracy or use of such information, or infringement of patents arising from such use or otherwise. Use of Microchip Technology products as critical components in life-support systems is not authorized except with express written approval by Microchip Technology Incorporated. No licenses are conveyed implicitly or otherwise under any intellectual property rights.

• Square 1 Electronics: "Easy PIC'n," by David Benson © 1997

• Dontronics: "Programming the PIC16F84," by Stan Ockers © 2000

CONTENTS

PREFACE

Over the years since Intel introduced its first microprocessor, literally thousands of microprocessors and microcontrollers have hit the market. Many were, and in some cases still are, very good devices. And, some were not so good. But, you would be hard-pressed to say that an immense variety of such ICs weren't available. That in itself is a good thing. Competition breeds both quality and value, and, in the end, the consumer reaps the benefit.

So, when Microchip Technology introduced the first of its peripheral interface controllers, there wasn't much fanfare over the event. At least, not at first! Once technicians, experimenters, and hobbyists became familiar with and began using these gems, however, enthusiasm soon hit a fever pitch. And a whole new chapter in the history of microprocessing was written.

Microchip's PICMICRO® MCU line is nothing short of miraculous in its performance and ease of use. Since the initial PIC16C54s, the company has produced and released an extensive list of devices to suit virtually any need and/or purpose. That, combined with the very simple 35-command instruction set, makes these integrated circuits perfect for many applications. Additionally, their low cost has made PICMICRO® MCUs attractive to the commercial electronics industry and others. Today, we see what seems like everything from hair dryers to toys to kitchen appliances being controlled by computer chips of one sort or another. I remember being astounded some years ago to find that my automatic dishwasher had an Intel 8085 running the show.

But, that is progress, and I doubt that manufacturers will ever look back. The microcontrollers just make things too easy and inexpensive for them to abandon this approach. Hence, microprocessors are here to stay. And, the better devices will be the ones to endure. All this brings us back to the Microchip line. The PICMICRO® MCUs have performed so well in so many different applications that they have literally become the front-runners in the microcontroller race.

Aside from price, there are a number of features of the PICMICRO® MCUs that catch the attention of designers and experimenters alike. Among these are the size and variety of chip packaging. This allows for very compact finished products that have the controller capability. Additionally, the "program on the fly" aspect of some of the ICs that permits the internal PICMICRO® MCU program to be changed from a remote location has been a big hit with industries such as cable television. This feature allows the companies to change access codes at will, ultimately curtailing much of the "pirating" of their signals.

In the final analysis, the PICMICRO® microcontrollers have garnered an intense following, as well as respect, among individuals that require computer control of a device. If you have something you need to regulate in some fashion, these little chips will likely do the trick. Of course, there are limitations regarding available memory space and so forth, but even those restrictions can, in most cases, be easily overcome. For example, additional memory space is handily accomplished with a number of small serial memory chips available at very reasonable cost.

Hence, I think you will enjoy this text. If you're like me, you would like to try numerous projects that can so easily be done with a microcontroller. In years past, I have either shelved or abandoned some efforts because of the tedium, complexity, or just plain inconvenience associated with many controller ICs. But with the PICMICRO® MCUs, I have been able to go back and complete some of those projects. And, that has made them very attractive to me. Hopefully, you, too, will experience that emotion. As they say, "Try it! You'll like it!"

CHAPTER 1
PICMICRO® MCUs

INTRODUCTION

In the past, I have found that the best place to start is at the beginning (I know, I know! I'm a master at stating the obvious). So, in keeping with that sage advice, let's take a look at the peripheral interface controllers themselves. In this fashion, you will become acquainted with these chips from the ground up.

As I stated in the Preface, Microchip Technology Inc. has produced an abundance of PICMICRO® MCUs over the years since the first ones hit the market. As of this writing, I counted 118 individual ICs in the company's inventory, with many more in the "planning" stage. These range from 8-pin to 84-pin devices and come in both Dual In-line Pin (DIP) and Plastic Leaded Chip Carrier (PLCC) configurations.

Additionally, some more unique arrangements, such as Plastic Quad Flatpacks (QFP), Plastic Shrink Small

With the variety of sizes, shapes, and lead layouts available, there is a PICMICRO® chip to satisfy virtually any need you might encounter.

Outline (SSOP), and Side Brazed Dual In-Line grace the ranks of PICMICRO® MCUs.

With the variety of sizes, shapes, and lead layouts available, there is a chip to satisfy virtually any need you might encounter. On that note, let the journey into the world of the PICMICRO® microcontroller begin!

PHYSICAL APPEARANCE

I have already given you a brief glimpse of what to expect, but allow me to go into more detail. In this way, you will know what is available from Microchip. This information will come in very handy when it comes time to think about the actual hardware arrangement of a prospective project.

DUAL IN-LINE PIN (DIP)

DIPs are probably the most familiar of all IC packages and fit neatly into special IC sockets.

As with all integrated circuits, the term "Dual In-Line Pin" refers to a rectangular device that has evenly spaced pins protruding downward from the long sides. These are probably the most familiar of all IC packages and fit neatly into special IC sockets. They can also be soldered directly to a printed circuit board, but I don't recommend that for experimental or hobby applications. The reason is that once soldered in, they are hard to get back off the board without damaging the board and the IC itself.

Anyway, the DIPs come in two varieties—plastic and ceramic. The ceramic versions are usually reserved for the Electrically Programmable Read Only Memory (EPROM) PICMICRO® MCUs, as these devices require a quartz window. The window exposes the internal silicon wafer allowing the user to shine ultraviolet (UV) light onto the wafer. Why do you want to shine UV light on the silicon? Good question! And, the answer is just as good! This is how you erase the memory. Why do you want to erase the memory? Well, at the risk of playing "20 questions," erasure is necessary when the program needs changing or doesn't work at all.

Hence, the EPROM PICMICRO® MCUs are used for prototyping where the code may have to be changed several times to "debug" it. Debugging refers to getting all the bad stuff fixed or out. (I hope this is making sense!) Ceramic is used in place of plastic, as it is more durable and better supports the quartz windows. But, it is more expensive! That, of course, means a higher price for the EPROM devices as opposed to the "One-Time Programmable" (OTP) chips. The latter is used for the finished project and holds the debugged software. You know, when everything is working correctly. OTPs are usually made of plastic (cheaper).

Okay, I've said a lot about the ceramic packages and not much about the plastic ones. Actually, there isn't a whole lot to say about the plastic cases. As stated, they are far less costly, and they are the most common encasements used in the manufacture of integrated circuits. If you work with ICs, then this is what the majority of the chips are housed in.

Now, we're talking about good plastic, not the cheap stuff. So, there is nothing wrong with using this material for the ICs. In fact, it makes perfect sense. The plastic just doesn't secure those little round quartz windows very well, so that is where the ceramic comes in. There is one other reason for the ceramic with EPROM chips. The lamps that produce the UV light for memory erasing tend to produce a substantial amount of heat, and the ceramic is better suited to handle that problem. Heck, the plastic might melt.

Oh, one other point. Whenever the chip itself produces substantial heat, as is the case with many microprocessors, ceramic is usually the choice. It's that same old "not-wanting-to-mess-with-hot-gooey-plastic" thing.

One problem with the DIP configuration is that it does restrict the number of pins the IC can have. Naturally, the more pins added, the longer, and bigger, the device becomes. This eventually leads to chips that are cumbersome, so other arrangements, which I will soon cover, take precedence. To date, 64 pins is the largest DIP I have encountered, and that one was really too large. So, as stated, the designers have come up with additional layouts.

PLASTIC LEADED CHIP CARRIER (PLCC)

In this version, the integrated circuit has leads on all four sides. The chip is usually square in shape, and the leads are not pins but metal contact points around the edges. If you have delved into computers at all, you have probably encountered these PLCC devices, especially with the older 286 machines.

This, of course, allows for more pins in a smaller space—an important point, as space, or the lack of it, is always a concern when prototype design is the issue. Manufacturers are striving to make everything from home computers to cell phones smaller, and this pin layout has helped. It does take a special socket, and if jarred too badly can let the chip become dislodged (hence, disconnected). But then again, this type of equipment is not supposed to be jarred that hard to start with.

The additional leads can be used for larger ports, added features, and a variety of other goodies. All of which makes the chip a more versatile item. So, it's easy to see how this design has garnered favor in the research and development (R&D) world, especially where size and space are important.

QUAD FLATPACKS

In this area, which is devoted to the ever intrepid "surface-mount technology" (SMT) realm, Microchip has introduced a mélange of devices for various applications. These include the Plastic Quad Flatpack (QFP), the Plastic Shrink Small Outline (SSOP), the Plastic Thin Quad Flatpack (TQFP), Plastic Small Outline (SOIC), and the Plastic Thin Shrink Small Outline (TSSOP). Each is designed to be soldered directly to the top of a printed circuit board. And, that saves space and eliminates some other problems.

In each case, the pins are bent at right angles to all four sides of the IC and lay flat against the surface they are set upon. These pins are then lined up and soldered to traces on the PCB. Hence the term "surface mount." This approach is rapidly becoming an industry standard, as it is fast and well suited to automated production. And that translates into *cheap*! I'm certain you have seen this type of construction in a number of products.

It's easy to see how the PLCC design has garnered favor in the research and development (R&D) world, especially where size and space are important.

One problem, or should I say step, these chips solve is the necessity of drilling holes through the printed circuit board. If the board is of the "double-sided" persuasion, then those holes also have to be "plated through." That is, they have to have a metal surface deposited on the inside of the hole itself to ensure a solid connection from one side of the board to the other. This is a time-consuming and non-cost-effective procedure that manufacturers love to avoid when possible. SMT has greatly reduced this requirement.

Additionally, surface-mount technology in general has allowed the overall circuits to be smaller. The ICs themselves are smaller, and due to the automated assembly, the boards can be much smaller. It would be redundant to say that this makes the companies building the equipment happy! But, it does make these designs difficult to work on. In fact, some of the discrete components, such as resistors, are approaching the size of a grain of sand. And, that makes them almost impossible to use for hobby construction.

As for the various package designations mentioned, the primary difference is in the thickness and size of the package. Essentially, the chips are functionally the same, just placed in modified enclosures. The purpose here is to give the manufacturing people variety, and they appreciate it. More than 100 million PIC16CXXX/PIC17CXXX devices ship each year and many are sent to original equipment manufacturers (OEMs).

So, these smaller SMT devices are becoming ever present in the world of electronics, much to the dismay of us experimenters and hobbyists. I mean, they—as well as their discrete cousins—are hard to work with. I have had more than one resistor or capacitor grasped in a set of tweezers, simply disappear into the ether, never to be heard from or seen again. You can look until you are age old and not find that little sucker. Somehow, they just instinctively know how to escape those tweezers.

That isn't meant to scare you! Just a little warning about working with SMT. If you do need compactness, surface mount is often worth looking into. But, personally, I try to avoid it as much as possible.

SIDE BRAZED DUAL IN-LINE

I have to admit that this one is a new one for me. I think I know what they are referring to, as I have seen components with the actual "pads" on the sides. Pads, that is, instead of any type of pin. But, to date, I have not employed such chips or, for that matter, components of this type, and have not seen an abundance of their use. That, of course, doesn't mean these devices are not highly useful. I'm sure in certain situations these gems really fill the bill. I just haven't been placed in any of those situations yet.

Anyway, be aware that this contact configuration does exist. It may well come in handy for you in the future. I mean, it may well come in handy for me in the future, but at the present time . . . well! Let me know if you find a really neat use for these.

As for the number of pins/contacts each chip utilizes, let me provide some examples that illustrate the pin arrangement of various PICMICRO® MCU types. As I said earlier, many of the chips you will be dealing with will have 18 pins. Some of these include the PIC16C52, PIC16C54, PIC16C58, and PIC16F84 (see *Figure 1.1*). Incidentally, the PIC16F84 contains an Electrically Erasable and Programmable Read Only Memory (EEPROM) that can be changed without UV erasure. This allows for the "program on the fly" approach and is a feature you will learn to love.

Examples of 28-pin PICMICRO® MCUs are the PIC16C56, PIC16C57, PIC16C62, PIC16C72, PIC16C76, and PIC14C000. This last one is used extensively in battery charging and maintaining. The PIC12C508 is an 8-pin device, while the PIC16C505 has 14 pins. The PIC16C64, PIC16C74, PIC16F874, and PIC17C44 are all 40-pin

**Figure 1.1.
A group of
PICMICRO®
microcontrollers
of various sizes,
styles, and pin
numbers.**

devices, with the PIC16C923 using 64 pins. The PIC17C752 and PIC17C756 are 68-pin chips and the PIC17C162 and PIC17C766 have 84 contacts.

So, that gives a small picture of the Microchip line of PICMICRO® microcontrollers. Each of the above ICs has its own features and characteristics, but that is part of what makes using PICMICRO® MCUs fun. I would be very surprised if you can't find something to fill your need.

Well, there you have it! Quite a selection of types, styles, and sizes of PICMICRO® microcontrollers. If you have a need, it is likely Microchip can help you. Next, let's briefly talk about the structure or architecture of these remarkable devices.

CHIP STRUCTURE OR ARCHITECTURE

In this section I will cover the basics regarding the structure used in the PICMICRO® MCU line. In the computer realm, they like to call this the "architecture," but what we will be discussing is the style of construction employed.

As you might expect, there have been a multitude of approaches used over the years. However, the Microchip selection is one that has a proven record.

REDUCED INSTRUCTION SET COMPUTING

As has been previously pointed out, the PICMICRO® MCUs use 35 instructions to perform their magic. Trust me, this is a small number of instructions as compared to many other microprocessor types. As the microprocessor

PICMICRO® devices use 35 instructions to perform their magic. Trust me, this is a small number of instructions as compared to many other microprocessor types.

evolved over the years, smaller instruction sets were a priority, but that was often difficult to achieve. There were just too many tasks to perform and all of them required a dedicated command.

Also, the physical structure of the IC played a role.Utilizing RISC architecture, Microchip was able to reduce the number of commands to the very manageable 35 instructions. And that makes them a joy to work with. If you appreciate nothing else about the PICMICRO® MCUs, I guarantee you will like the instruction set. Especially if you have worked with some of the other microprocessor and/or microcontrollers.

On a last note regarding chip construction, the PICMICRO® microcontrollers use the Complementary Metal Oxide Semiconductor (CMOS) approach, and that ensures stable operation and low power requirements, yet another attractive feature for the OEM.

As I said, this is a brief look at architecture. If you want more detailed information, check the Microchip web site, and there are many books available that will deeply delve into this. However, it is not information that you need in order to use the PICMICRO® microcontrollers.

HARDWARE REQUIREMENTS

Here, we will take a gander at the support hardware and tools that help you develop your projects. This is a combination of hardware and software, all designed to put your program into a PICMICRO® MCU.

THE PERIPHERAL INTERFACE CONTROLLER PROGRAMMER...

I'll go into the theory of these programmers in much more detail in Chapter 3, but just for fun, let me cover the basics here.

...Actually, that should read "Programmers"—you know, plural, as there are dozens of them available commercially and on the Internet. These are usually fairly simple devices that utilize your home computer's parallel port (although some use the serial port) and are quite simple to use.

In Chapter 3, a very simple system is described to program the venerable PIC16F84 chip, an excellent example of a typical PICMICRO® MCU programmer. When combined with some basic software, the device will "burn" into its memory any program that will fit in the chip's memory. *Voila*!

I'll go into the theory of these programmers in much more detail in Chapter 3, but just for fun, let me cover the basics here. An archetypal programmer consists of two sections—the power supply and the control unit. These two work together, along with the software, to achieve programming of the PICMICRO® MCUs.

The power supply has to deliver two separate voltages. One is the "chip operating" voltage, usually 5 volts, and the other is the "program" voltage, somewhere between 12 and 14 volts (usually 13.5). These voltages are normally provided by linear regulators and maintain a stable level. As you might expect, the program voltage is only used to actually burn the program into memory. It is the only occasion I can think of where you would want to expose a PICMICRO® MCU to 13.5 volts of juice.

The control section is connected to the host computer port via a standard cable and receives the installation commands and software that will be put into memory. Here, the "host" computer refers to your PC and the port refers to either the serial or parallel port of that PC. This segment of the programmer normally consists of a buffer IC, such as the 7407, and some transistor switches. Of course, there will also be a socket to plug the PICMICRO® MCU into.

Most programmers include a light-emitting diode (LED) to indicate when the unit is operational. They also often include a "header" that allows larger sockets to be connected to the system. These sockets accommodate PICMICRO® MCUs with 28 or 40 pins, and some programmers can even handle 64-, 68-, or 84-pin devices through external sockets. They are powered by either a DC "wallwart"-style supply or batteries. The battery-operated scenario can be handy for portable use, but the DC supply will normally be more reliable. Whichever way you go, the input DC must be high enough for the regulators to produce the program voltage.

The market is full of very dependable programmers from a number of manufacturers, including Microchip (see the Source List in the Appendix), or you can opt to build your own. The commercial units range from dedicated devices for, say, the PIC16F84 to universal programmers that handle the entire Microchip line. Naturally, the amount you want to spend will dictate how complex a programmer you will receive.

One unit that I bought when I first became interested in this subject is the EPIC Plus Pocket PICMICRO® Programmer (see *Figure 1.2*) by microEngineering Labs, Inc. This is a device that will program many of the PICMICRO® MCUs, has the external socket header, and has served me well. I still use it often. One drawback is that it will not program the PIC16C54. But then, my favorite is the PIC16F84, so, in my case, this is not a really significant deficiency.

Figure 1.2. microEngineering Labs' EPIC Plus Pocket Programmer pictured with a group of PICMICRO®s.

A quick caution here: once you get to playin' with these PICMICRO® MCUs, you are likely to find that you want to expand into a wider range of devices.

That's just one of many, many programmers available, however, and most are sold at reasonable prices. Another factor that should help guide your decision—and purchase—is how far you plan to go down this road. A quick caution here: once you get to playin' with these PICMICRO® MCUs, you are likely to find that you want to expand into a wider range of devices. Hence, a slightly more versatile, and expensive, programmer might well be a prudent choice.

DEVELOPMENT SOFTWARE

To make working with the PICMICRO® MCUs even easier, Microchip provides an excellent group of software it calls the MPLAB® integrated design environment (IDE). The latest version is 4.12.12 and can be downloaded from the Internet or obtained free from Microchip Technology. And, this is the proverbial "whole ball of wax."

MPLAB® IDE contains a text editor, assembler (MPASM™ assembler), and device programmer (PICSTART® Plus), and this stuff will run under DOS or Windows. These programming tools can be used with just about every commercial programmer to create the assembler code and assembled HEX file. At that point, however, to load the code, it will be necessary to change to the software for your particular programmer.

The nice part about MPLAB® IDE is that all the needed programs are available in one place. Simple clicks of the mouse will take you to wherever you need to be. And, that makes for a very simple operation. It also makes your life easier! The MPASM™ assembler program is available separately for DOS, but I would highly recommend getting MPLAB® IDE and keep it all in one program.

One really convenient aspect of all this software is that Microchip is constantly updating it (the DOS versions are not being updated anymore). And, those updates are available from Microchip either by mail or on the Internet. In that fashion, your development tools are always current. One last comment regarding the PICSTART® Plus software: it is designed to accommodate the entire PICMICRO® microcontroller line and updates will maintain that status for all new devices. However, it can only be used with Microchip's PICSTART® Plus programmer.

SOFTWARE REQUIREMENTS

With the MPLAB® IDE in hand, you will be able to write the program using the included text editor, then assemble it with the MPASM™ assembler program, also included within MPLAB® IDE. Once this is done, the PICSTART® Plus program will install the project software in the PICMICRO® microcontroller if you are using the Microchip PICSTART® Plus programmer. This is all very handy to have in one package, but if you are using a different programmer, it will be necessary to leave the MPLAB® IDE after assembling the program and install it using the software provided by the company that makes your programmer.

If that sounds confusing, let me elaborate. The MPASM™ assembler will develop the code format (HEX) that the PICMICRO® MCUs want, and that code can then be burned into a PICMICRO® microcontroller with a programmer. However, this step must be done with the proper software designed for the particular programmer. If you are using Microchip's programmer, you're in business, as the proper software is included in MPLAB® IDE. If not, the drivers for whatever programmer being used must also be employed. I hope that clears things up without being redundant or more confusing. The text editor in MPLAB® IDE will allow you to write the program in an understandable format. This program is really just a simple word processor that produces text the assembler will comprehend. You will be able to add notations and instructions on the text for both explanation and direction, and these will not interfere with assembly of the code. This is done by using a semicolon before anything that is not actually part of the code. But, don't worry too much about that right now, as it will be explained in greater detail in the software chapter (Chapter 11). Of course, you may use assemblers other than the MPASM™ assembler, but there are some good reasons not to. First is compatibility! The MPASM™ assembler is used almost universally. Hence, you won't have to learn more than one method. Also, it is a Microchip product, so it will be compatible with Microchip ICs. And, it is free and available from the company. Actually, I could come up with other reasons, but those should be sufficient. Trust me, it is always a good idea to take advantage of all the compatibility you can lay your hands on when it comes to working with computers.

As I said a moment ago, the use of the MPLAB® IDE and MPASM™ assembler will be covered in more detail in Chapter 11, but I am trying to get you acquainted with some of this stuff now. This is a very simple procedure compared to the majority of microcontrollers, and I think you will enjoy working with these chips once you get used to them. That is especially true if you have had any experience with some of the others.

As the last part of Software 101, let's look at the instruction set. The 35 instructions are arranged in categories according to the function they accomplish. So, look them over and again, much more will be said about these commands. Actually, you will notice that this listing contains 37 instructions. Two are obsolete, but everyone I have talked to says to "keep them," as they make things easier. And that seems to be true.

MICROCHIP PICMICRO® INSTRUCTION SET

CHANGE REGISTER CONTENTS:

CLFR	f	Sets all bits in f to 0
CLRW		Clears W register to 0
COMF	f,d	Complements selected register, swaps values, result in W or f
DECF	f,d	Decrements the selected register, result in W or f
INCF	f,d	Increments the selected register, result in W or f
BCF	f,b	Sets bit b in register f to 0
BSF	f,b	Sets bit b in register f to 1
RLF	f,d	Rotates bit in f one position to left, bit rotate through carry
RRF	f,d	Rotates bit ib f one position to right, bit rotates through carry
SWAPF	f,d	Swaps most significant and least significant nibbles of selected register

As the last part of Software 101, let's look at the instruction set. The 35 instructions are arranged in categories according to the function they accomplish.

MOVE OR DEFINE DATA:

MOVLW   k   Loads W with literal
MOVF   f,d   Moves selected register into W or f
MOVWF f   Moves W into selected register

CONTROLLING PROGRAM FLOW:

CALL   k   Call subroutine
RETURN   Return from call
RETLW   k   Return from call and load literal k into W
RETFIE   Return from interrupt
GOTO   k   Goto a specific address
BTFSC   f,b   Test b bit, register f. If bit is 0, skip next instruction
BTFSS   f,b   Test b bit, register f. If bit is 1, skip next instruction
DECFSZ f,d   Put f-1 in register W or f. If result is 0, skip next instruction
INCFSZ f,d   Put f+1 in register W or f. If result is 1, skip next instruction

CONTROL MICROCONTROLLER:

TRIS   f   W bit determines port line input/output on line-by-line basis
SLEEP   For power consumption. Microcontroller in sleep mode
CLRWDT   Clear watchdog timer and resets prescaler
OPTION   W contents sent to option register

Most of the instructions practically spell themselves out. Once you become familiar with them, writing software will be a breeze.

LOGIC:

ANDLW	k	Place AND contents of W with literal in register W
ANDWF	f,d	Place AND contents of W in register W or f
IORLW	k	Place OR contents of W with literal in register W
IORWF	f,d	Place OR contents of W in register W or f
XORLW	k	Place XOR content of W with literal in W
XORWF	f,d	Place XOR content of W in register W or f

ARITHMETIC:

SUBLW	k	Subtract W from literal and place result in W
SUBWF	f,d	Subtract W from f and place results in d
ADDLW	k	Add literal to W and place result in W
ADDWF	f,d	Add W to selected register and place result in W or f

NOTHING:

NOP	Do nothing for one instruction cycle (time delay)

And there is the entire instruction set including the two obsolete commands. As can be seen, this is easy stuff in terms of microcontroller programming.

Of course, these instructions will be covered in far more detail in later chapters, but, even at a quick glance, it is more than apparent how simple this approach can be. Most of the instructions practically spell themselves out. Once you become familiar with them, writing software will be a breeze.

CONCLUSION

Well, that should provide a solid basic introduction to PICMICRO® microcontrollers. All in all, it doesn't get much simpler than this. And the results will delight you! So many tasks that previously had to be done with dedicated integrated circuits can now be done by putting a short program into one of the many PICMICRO® MCUs available. This not only makes the process of developing projects easier, it insures that you get exactly what you want. And, that is a feature that can well be worth its weight in gold. From here on out, you will be able to design devices that perform the job you had in mind.

So many tasks that previously had to be done with dedicated integrated circuits can now be done by putting a short program into one of the many PICMICRO® MCUs available.

CHAPTER 2
A PICMICRO® MCU PROJECT PROTOTYPE BOARD

INTRODUCTION

When approaching a microcontroller project, you have to move in one of two directions: Either you develop the software first, then the hardware, or vice versa. This is akin to the chicken-or-the-egg debate that is age old. In the end, however, what you start with isn't going to make a whole lot of difference, as you are going to need both if the project is to be successful.

On that note (or perhaps bit of philosophy), let me introduce you to the hardware-first approach. I stress that this is merely a personal preference on my part, and the other path will work just as well. But, for the purpose of this book, let's start with the prototyping board and its construction.

THEORY

With this piece of "hardware" in hand, you will have the tool needed to lay out the projects and test the software. It

With this piece of "hardware" in hand, you will have the tool needed to lay out the projects and test the software.

may look a little complex at first, but as I go through and explain each section, you will see that it is a simple, but effective, way to work with PICMICRO® microcontrollers.

Referring to the schematic in *Figure 2.1*, let's start with the Dual In-Line Pin (DIP) switches S3 to S9 and S29 to S35. These are found in the main body of the diagram. Note that each switch is connected to a 10,000-ohm resistor that is, in turn, connected to the positive 5-volt rail. Further note that the other side of each switch is hooked to a Single In-Line Pin (SIP) socket. The purpose of the sockets will be explained in a moment, but first, let's look at the switches and resistors themselves.

When working with micro-controllers—or, for that matter, most digital chips—it is rarely advantageous to leave any logic pin (line) uncon-nected or "floating."

When working with microcontrollers—or, for that matter, most digital chips—it is rarely advantageous to leave any logic pin (line) unconnected or "floating." This can cause unwanted interference and excessive power consumption. So, to get around this problem, "pulling" those otherwise unconnected pins high (positive) is a commonly used remedy. Of course, that's where the switches and resistors come into play. It is *not* recommended that you apply the full positive potential to the line—hence, the 10,000-ohm resistors are used to "drop" that potential to a safe level.

Now, when the need arises, you will be able to apply a positive potential (pull the pin high) to those logic pins that will not be used. Thus, these switches provide the versatility and convenience that any good prototype lab should have.

Moving toward the center of the schematic (the ZIF socket), we come back to the SIP sockets (SIP1 and

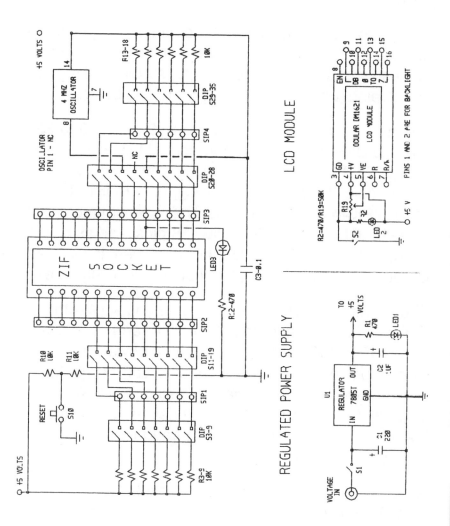

Figure 2.1.
Prototype
lab schematic.

SIP4). As promised, I will explain their presence on the board. As a matter of fact, this explanation will serve for all the SIP sockets—that is, they furnish "convenient" access to various points and/or components on the lab. In this fashion, you will be able to bypass the pull-up switches and resistors and go directly to the PICMICRO® microcontroller in question. At the surface, these may seem excessive, but once you begin working with the prototype board, you will soon recognize the expedience of these sockets. And, as previously stated, making your prototyping job easier is a big part of this board.

The next sets of switches, S11 to S19 and S20 to S28, are also DIP switches, and they control the access to the PICMICRO® MCU you are working with. In this way, you will be able to apply whatever signal is required to the various PICMICRO® MCU pins. The most commonly employed PICMICRO® MCUs have 18 pins, and as can be seen, that is how the board is laid out. There is access to larger PICMICRO® MCUs, but that will be covered next.

These switches will be most useful as you work with your projects. For example, the PIC16F84 (and most 18-pin PICMICRO® MCUs) has the positive power on pin 14. Hence, power to the microcontroller can be controlled with the switch that is connected between pin 14 and the +5 volts. Additionally, the clock signal, as well as the actual data, can be manipulated through the switches. I think you will like this part.

Next, we find some more SIP sockets and the Zero Insertion Force (ZIF) socket. Let me first address the ZIF socket. Since the PICMICRO® MCU will probably have to make many trips between the prototype board and the programmer, a Zero Insertion Force socket is a very pragmatic device. If you have worked with integrated circuits (ICs) at all, then you are all too familiar with those very delicate pins that protrude from the sides of these devices. They have a nasty habit of bending when you try to install them in the traditional friction grip-style IC sockets, and if bent back and forth very often, they just plain break off. This is not good! You have spent your hard-earned money on a PICMICRO® microcontroller, and suddenly one or more of its pins break off. Curses!

There is an answer, however, and it is the ZIF socket. Here, a lever opens the socket, you simply drop the PICMICRO® MCU into it, and the lever then closes the socket. Nice and neat! Now you will be able to plug said PICMICRO® MCU into and out of the socket as often as necessary without the danger of bending and/or breaking the pins from having to force the chip in, or pry it back out.

Now, since there are a number of PICMICRO® MCUs that have more than 18 pins, and some that have less, we need a ZIF socket that will accommodate this variety. ZIF sockets generally come in two types—dedicated and universal. The dedicated devices are made for chips with a specific number of pins and a specified pin spacing. For example, 28-pin ICs traditionally have 0.6 inches of space between the pin rows, while 18-pin devices have 0.3-inch spacing. For our purpose, those don't work.

Never fear, though! The universal socket allows for both 0.6- and 0.3-inch spacing, as the socket's slots are over-sized. This is our boy! With this type of socket, PICMICRO® MCUs from eight to 28 pins will fit the proto-type board. There are some 40-pin PICMICRO® MCUs, but they are beyond the scope of my ability and this book. I seriously doubt you will work with very many of them, at least not at first anyway.

Normally, you will work mostly with the 18-pin beauties, so I designed the primary circuitry of the board to handle

Those very delicate pins that protrude from the sides of these devices have a habit of bending when you try to install them in, and if bent back and forth very often, they just plain break off.

these. But, I did include SIP sockets that access the remaining 10 pins for the larger 28-pin chips, if you do work with some of those. These sockets are part of the SIP rows SIP2 and SIP3. Additionally, both rows provide direct access to all 28 pins of the ZIF socket. In this fashion, you will be able to bypass all switches and resistors, etc., if you so desire. This option will come in handy if you use the breadboard for the majority of the project wiring, or if you do work with a large-size PICMICRO® MCU.

To finish off the main body of the prototype lab, a socket is provided in the upper right hand area for a crystal oscillator. I find these are the best, and most reliable, way to provide the necessary clock for the PICMICRO® MCUs. Many of the chips, such as the PIC16C54 and PIC16F84, run on a 4-megahertz clock, but some versions of the microprocessors can use a 10-megahertz oscillator. Hence, a socket that allows the crystal oscillator to be changed is an attractive feature.

The light-emitting diode/resistor network (LED3/R12) is included to indicate when 5 volts is applied to the ZIF socket's pin 14. As I said earlier, there may be times when you will want to cut power to the PICMICRO® microcontroller and this indicator will let you know the power status. Also, C3 is included as a "bypass" capacitor. This is a commonly used technique in digital circuits to provide a little reserve power when the main power is interrupted. It will help eliminate glitches.

Last, but as they say, certainly not least, is the S10/R10/R11 network. This is what is affectionately called, in computer circles, the reset. Its job, as the name implies, is to reset the entire system when a major problem or glitch is encountered. You know, when the whole thing locks up and won't do anything else. I trust everyone has been through that situation at least once.

Reset with the PICMICRO® MCUs, as with many micro-processors, is accomplished by placing a "low" signal on the reset pin. In our setup, R10 keeps that pin "high" until switch S10 is pressed. That action places the requisite low on the pin, and the chip resets. Trust me, you will find this little button most useful during your adventures with PICMICRO® microcontrollers.

And that about covers the principal sections of the pro-totype lab. But that's not all there is. So, let's look at the remaining goodies included on the board designed to make your "PICing" a happy experience.

As you are probably aware, many digital integrated cir-cuits are quite particular when it comes to the voltage you apply. PICMICRO® microcontrollers are no excep-tion. They like positive 5 volts and not much else. Actu-ally, that's a misstatement. They like positive 5 volts and don't—absolutely don't—like anything else. In fact, you run the serious risk of burning them up if you apply more than 5 volts, and if that voltage is much lower than 5, they often perform erratically.

Hence, it is very prudent to provide a nice stable source of positive 5 volts. That statement should immediately bring a certain IC to mind—namely, the voltage regulator! That's right, the very first integrated circuit depicted on the schematic is our old friend, the LM7805. This device will limit the voltage applied to the rest of the circuit to a value very close to, if not exactly, 5 volts. That level is also quite important with the LCD module that I will cover in a moment.

The additional components for this "power supply" are as follows. C1 acts as the input voltage filter, C2 as the output voltage filter and bypass unit, and the LED1/R1 network is the "power on" indicator. Naturally, the negative side of the input voltage source is connected to the prototype board's ground rail.

All of this is a very common linear regulator circuit that I trust you have seen and used in the past. While not the most efficient way to do the job, it will provide a very stable system voltage that protects your valuable prototype board.

The next area of interest is depicted at the top of the diagram. Here we are employing a liquid crystal display (LCD) module as an accessory to the prototype lab. The LCD module I used is an Ocular OM1621, but any module that employs the Hitachi HD44780 controller chip or equivalent will do the job. And, that is a good number of these modules. The main reason I used the Ocular device is that I had one on hand, and I'm a firm believer in not buying anything you already have when building these projects. I mean, there's no sense in making this prototype lab any more expensive than necessary.

Anyway, LCD modules are a great way to go, as they include most of the electronics needed to get the LCD display working. If you have tried to build projects that utilize a separate controller and LCD display, you will appreciate these modules. And they have become abundant and *cheap* on the surplus market (see the Source List in the Appendix).

SEMICONDUCTORS:
7805T Positive Voltage Regulator (1)
Red Light-Emitting Diode (1)
Green Light-Emitting Diode (1)
Orange Light-Emitting Diode (1)
4-Megahertz Crystal Oscillator (1)
Ocular DM1621 LCD Module (1)

RESISTORS:
10,000 Ohm ¼ Watt (14)
470 Ohm ¼ Watt (3)
100 Ohm ¼ Watt (1)
50,000-Ohm Potentiometer (1)

CAPACITORS:
0.1-Microfarad Monolythic (1)
1-Microfarad Electrolytic (1)
220-Microfarad Electrolytic (1)

SWITCHES:
SPST Slide Switch (1)
9-Position DIP Switch (2)
8-Position DIP Switch (2)
Momentary Push Button
Switch (1)
SPST Rocker Switch (1)

SOCKETS:
SIP Sockets (41)
28-Pin DIP Socket (1)
28-Pin ZIF Socket (1)
14-Pin Oscillator Socket
(Optional)

**OTHER
COMPONENTS:**
Coaxial Power Jack (1)
PCB Materials
Hardware,
Solder, Wire, etc.

*Table 2.1.
Prototype
lab parts list.*

Associated with the module are a couple of other components. Nothing fancy, just a switch (S2) that applies power to the system, an LED/resistor network (LED2/R2) that tells you power is "on," and a potentiometer (R19) that adjusts the LCD display contrast. The modules have a "VEE" pin on them that controls the contrast of the display, and the potentiometer allows the voltage on the VEE pin to be varied to whatever level provides the desired contrast. That can be from solid blocks to nothing at all. So, adjusting the contrast is important in having the display function the way you want it to function. And, if you haven't had experience with these modules, you will discover that fact as you work with them.

The last addition to this section of the lab is more SIP sockets. As with the main area of the board, these are used to provide access to the various module pins (most modules have 14, and I will cover that next). With the sockets, you will be able to address just the module inputs/outputs that are needed.

These modules are not very complex in nature. As with much of today's electronics, the chips do all the work.

So, let's take a quick and dirty look at what the module pins are used for. Bear in mind that this is a somewhat "generic" description, and you may encounter different layouts with different modules. It's always best to obtain the company data sheets if possible. But, the vast majority of the units using the 44780 chip also use this wiring scheme.

Pin 1 is ground, Pin 2 is VCC or the positive power supply, and Pin 3 is VEE or contrast adjustment. As promised, a quick note here about LCD module voltages. As with the PICMICRO® MCUs, they like +5 volts and will burn quickly if much more than that is applied. So, be very careful with the modules. Pins 4, 5, and 6 are the control inputs, Register Select (RS), Read/Write Select Signal (R/W), and Operation (E, or data Read/Write enable signal), respectively. And finally, Pins 7 through 14 are the eight data ports DB0 to DB7, with 7 being the least significant bit (LSB) and 14 being the most significant bit (MSB). So, as can be seen, these modules are not very complex in nature. As with much of today's electronics, the chips do all the work.

The schematic diagram clearly illustrates how the voltage pins (1, 2, and 3) are connected, and the PICMICRO® microcontroller/firmware will handle the other functions— that is, when the RS, R/W, or E signals are needed and what goes on the data bus (DB0-DB7). You will find with many projects only the DB4 through DB7 data pins are employed, as these are the only ones necessary.

In addition to all of what I have just covered, a few related items are available with the lab: a solderless breadboard for laying out your support components, a 16-key keypad (Is that redundant? Nah, I don't think so.) that will be needed for some projects, and a four-digit multiplexed light-emitting diode (LED) display for devices that use such displays. As you work through the various projects described in this book, you will be using all of these "extras." And, for convenience, both the keypad and the LED display employ—you guessed it—SIP sockets!

With everything in one place, your lab is a self-contained unit. And, that provides a certain degree of convenience. Hopefully, it will become a good friend as you make your way into the world of PICMICRO® microcontrollers. I know mine has! If you are already in that world, I think this lab will provide a valuable tool.

CONSTRUCTION

Putting this gem together is really very simple. I laid it out on a 6½-inch by 9½-inch piece of quarter-inch Plexiglas, but other plastics, wood, or even metal (if you are careful about shorts) will do just as well. Also, my choice of section placement could easily be varied to suit your taste. All in all, there is nothing particularly critical about this piece of gear.

For the main section of the lab, I used a printed circuit board (PCB) pattern. This is not etched in stone, however, and perf-board/point-to-point wiring is very definitely an option. The bottom line here is determined your capabilities and by what materials you have on hand.

I found it best to install the components from the center of the PCB out. That helped keep the congestion down. The 28-pin friction-grip socket used for the ZIF socket went on first. It is best to use this arrangement, as it would be hard to remove the expensive ZIF socket if it is soldered directly to the board.

Following that, the SIP sockets went on, then the resistors. Now the DIP switches, push-button switch, and other components are soldered to the board. The last areas I worked on were the regulated power supply, the clock, and finally the larger power switches.

With everything in place, a quick check of the work was in order. Once I was satisfied that there were no mistakes, I ran a check on the voltage. When that proved to be correct, this section was ready to go.

All segments of the lab are secured to the Plexiglas with bolts and standoffs. I like to use plastic beads that can be bought at hobby supply centers for standoffs, as they are easy to use and low in cost. This approach makes it simple to change things if the need arises. Also, it's easy to remove sections that might need repair or modification.

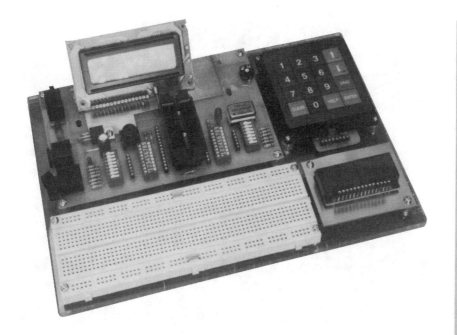

Figure 2.2.
The completed
prototype lab.
Note the keypad,
LED display,
breadboard, and
LCD module
built in for
convenience.

Other than placing some self-adhesive rubber feet on the bottom as bumpers, that's about all there is to building the prototype lab. A couple of nights' work, and it will be up and ready to go to work for you.

CONCLUSION

With the prototype lab completed (see *Figures 2.2* and *2.3*), you will be able to whip through those projects in no time at all. This, of course, makes the experience of working with PICMICRO® MCUs far more pleasant. Hence, this is a good place to start! The ability to lay out your projects before writing the software will prove to be a pragmatic approach. In this fashion, the ports used,

Figure 2.3.
A close up of the actual electronics of the prototype lab. The regulated power supply is located in the upper left section of this board, and the crystal oscillator is to the right.

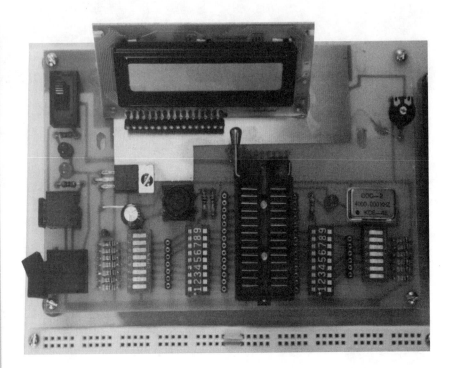

the chip clock speed, and other such details will be readily at hand when it comes time to write the code.

So, have fun with this and use it well. As I said earlier, it will probably become one of your best friends when dealing with PICMICRO® microcontrollers. At least, it will make life a little easier!

CHAPTER 3

A PIC16CXXX/ PIC16F84 PC-BASED DEVELOPMENT PROGRAMMER

INTRODUCTION

In this chapter, we will examine a simple, but effective pro-
grammer for the PIC16CXXX/PIC16F84 microcontrollers.
The information comes from Microchip's application note
AN589 (*Figure 3.2*), which is authored by Robert Spur of
Analog Design Specialist, Inc. The finished product will
allow you to program the very popular and versatile
PIC16CXXX/PIC16F84 line of MCUs.

These integrated circuits have become popular due to
their internal Electrically Erasable and Programmable
Read Only Memory (EEPROM). This feature allows the
user to change the memory content without the necessity of ultraviolet (UV) light erasure, which is not only a
cumbersome but also a time-consuming process.

Additionally, these microcontrollers can be reprogrammed
while in their prospective circuits. That is, the chip doesn't
have to be removed from the device it controls to be

*The finished
product will
allow you to
program the
very popular
and versatile
PIC16CXXX/
PIC16F84 line
of MCUs.*

This programmer can easily be incorporated into any project where you might want program-on-the-fly capabilities.

programmed. That leaves a multitude of possibilities open. For example, control systems for manufacturing, medical equipment, traffic management, or other situations can easily be changed through modem access or popping a disk into a connected drive.

An industry that has made good use of this type of programming is cable television. With piracy being a major problem in this area, the various cable companies can periodically, and at will, reprogram the individual reception "boxes." The process is simple, effective, and not terribly painfully to subscribers. Also, it does keep the hackers out of the system, at least for a while. They will eventually get back in, but then the cable company can again change the control code.

So, the PIC16CXXX/PIC16F84 ICs have gained a lot of support among designers and hobbyists alike. And the good news is this programmer can easily be incorporated into any project where you might want program-on-the-fly capabilities. Incidentally, I'm referring to these microcontrollers as PIC16CXXX/PIC16F84 because both the "C" (CMOS) and "F" (Flash) versions still exist as of this writing. However, the "C" chips are being phased out in favor of the flash-style devices.

Furthermore, this system is capable of reading or verifying data from a remote unit. Again, the microcontroller does not have to be removed from the existing equipment to perform this task. And the remote chip can be updated with ease in the event minor changes have to be made to the code.

Well, that should partially illustrate how useful the PIC16CXXX/PIC16F84 ICs can be, but there is much more to this story. Once you begin working with these chips, I suspect they will become one of your favorites. They sure are for me!

On that note, let me describe the programmer. First, we will look at how the device works, then take a look at the operational theory. Buckle up, because here we go!

THE CIRCUIT

For this section, we will be referring to *Figure 3.1*, the schematic diagram. This device provides "serial stream" programming of the PICMICRO® microcontroller via a parallel connection (Centronix) to the host personal

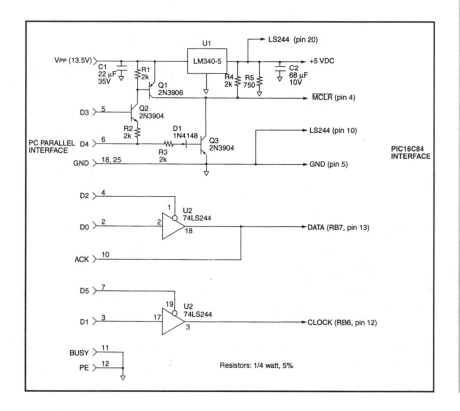

Figure 3.1. Schematic diagram of the PIC16F84 PC-based programmer.

computer. However, a microprocessor parallel interface can be achieved by substituting an "output command" for the "biosprint" command in the code.

Regarding the mechanics of the unit, Q3 is turned on via a "high" signal on D4 of the PC parallel interface. This results in the "master clear" (MCLR) pin going low, which resets the PIC16CXXX/PIC16F84. Next, the reset condition is terminated, and the "program/verify" voltage is applied to the MCLR. This is usually in the 12- to 14-volt range, with 13.5 volts the norm. A high on D3 and a low on D4 accomplishes this by turning Q3 off and Q1 and Q2 on.

Note that circuit protection of Q1 and Q3 comes from connecting the emitter of Q2 to D4. This action prevents reset and program/verify from occurring concurrently. Also, the 2N3904 used for Q2 has a 6-volt breakdown that is not exceeded by the normal operational voltage (5 volts).

Notice resistors R1, R2, R3, and diode D1. These components comprise a logic level to analog interface. R4 is included as a pull-up resistor that provides a logic high on the MCLR during the "run" mode. This will be covered further in the next section.

R5 is used to maintain proper voltage regulation during the program/verify cycles. In that mode, 13.5 volts are applied to the output of U1, the LM340-5 voltage regulator, through resistor R4, and R5 will preserve the positive 5 volts needed by the PIC16CXXX/PIC16F84. By the way, the 13.5 program/verify voltage does have to come from a regulated external source.

U2 is a "tri-state" buffer that controls the data flow and clocking. As seen in Figure 3.1, data is handled by the parallel PC connections to D0 and D2, while the clock is managed by D1 and D5. The tri-state buffers, in addition to buffering a line, function like switches. Hence, each buffer has to be turned on in order to affect a signal throughput. This is done by D2 and D5 for data and clock, respectively.

During programming, D2 and D5 enable both buffers, allowing data and clock signals to be transferred from the PC to the programmer. This, of course, facilitates the transport of information into the microcontroller. During "read," the printer data acknowledgment line (ACK) is used to send verification info back to the PC. This is necessary because the data buffer is tri-stated, or turned off, by D2 during a read cycle.

When you're done either programming or reading the PICMICRO® MCU, D2 and D5 tri-state both buffers, effectively isolating the programmer from the target circuit. The purpose here is to provide the target access to the various PC parallel lines without interference from the programmer. At this point, the programmer remains physically, but not electrically, connected to the target circuit.

The verification process is as with most microcontrollers, first setting the supply voltage to the maximum recommended level, then setting it to the minimum recommended value. Each time, a program/data read should be performed to confirm the validity of the internal data.

All right! That pretty well describes the hardware and operation of the programmer. As with so much associated with the Microchip integrated circuits, this device is very straightforward. Next, let's look at the functional side of the process.

During programming, D2 and D5 enable both buffers, allowing data and clock signals to be transferred from the PC to the programmer.

THEORY

In this section, operation of the programmer, as dictated by the software, will be the primary point of concern. We now know how the hardware works, but it does need commands to do its job. So, circuit management will be examined here.

In order to program a PIC16CXXX/PIC16F84, you have to put the microcontroller in the program mode. I know, that's a rather simplistic statement, but in its defense, it is also an accurate statement. To do this, a low must be placed on the RB7 (pin 13) and RB6 (pin 12) lines, while pin 4 (MCLR) is first brought low, then brought to the 13.5 program voltage (actually, that can be between 12 and 14 volts).

The low-to-high transition of the master clear (MCLR) does just that—clears the PICMICRO® MCU. That is, setting pin 4 low resets the chip, and bringing it high readies the PICMICRO® MCU for programming.

The MCLR will have to stay at that voltage until programming and verification have been completed. Data entry is done via the RB7 line, in which the information is entered into the microcontroller in a serial fashion. A high-to-low change on the clock line (RB6) after each bit of data qualifies said data.

The command field is comprised of the first six bits, while the last 16 bits form the data field. This data field consists of a zero starting bit, 14 true data bits, and another zero stop bit. This serves to differentiate the information being entered into the device.

Following program/verify, the MCLR is brought low to reset the PICMICRO® microcontroller and is then electrically released. The purpose here is to free the master clear for use by the target circuit. R4 is in place in case the target doesn't bring MCLR to the high status. In that scenario, the pull-up resistor will "pull" pin 4 high. This is important, as it facilitates microcontroller program execution unconstrained by connection to the programmer.

In the read mode, following the six-bit command field, the direction of data flow from RB7 is reversed. This permits the data to return to the programmer. As previously mentioned, in the read mode, the data buffer is tri-stated (turned off), and this allows return of the requested data to the programmer.

RB6 plays a role in all this by utilizing the rising and falling edges of the clock pulses. The rising edge moves a bit out of the PICMICRO® MCU to the programmer, and the falling edge of each pulse qualifies the bit.

Hence, it takes 16 clock cycles to move the 14 "true" data bits. To prevent an accidental programming cycle during read, the MCLR voltage is never taken above the maximum operational voltage (6 volts). This also prohibits that voltage from exceeding the maximum levels of RB6 and RB7.

A last notation from Microchip is that during programming, provisions should be made to keep the target circuit from resetting the target microcontroller or affecting the status of RB6 and RB7. However, this is accommodated with the hardware/software arrangement of this programmer.

SOME SOFTWARE NOTES

Sending new data to, or reading data from, a remote system can be done by reading the file from a standard PC PROM device (i.e., a floppy or hard disk) and calling the appropriate function in *Figure 3.3.* By using the proper command name and providing the data to be transferred, the target circuit will receive the firmware update. For convenience, the command names are listed below.

LOAD_CONFIG sets PIC16CXXX/PIC16F84 data pointer to configuration.
LOAD_DATA loads, but does not program, data.
READ_DATA reads data at current pointer location.

Figure 3.3.
Software
examples for
managing target
systems.

```
EXAMPLE 1:   PUT TARGET SYSTEM INTO PROGRAM MODE
    .. program code..
    ser_pic16c84(PROGRAM_MODE,0);
    .. program code..

EXAMPLE 2:   READ DATA FROM THE TARGET SYSTEM
    .. program code..
    data = ser_pic16c84(READ_DATA,0); // read data
    // transfers data from target part to variable "data"
    .. more program code..

EXAMPLE 3:   PROGRAM DATA INTO THE TARGET SYSTEM
    .. program code..
    ser_pic16c84(LOAD_DATA,data);// load data into target
    ser_pic16c84(BEGIN_PROG,0);// program loaded data
    ser_pic16c84(INC_ADDR,0);// increment to next address
     // transfers data from program variable "data" to target part
    .. more program code..

EXAMPLE 4:   PUT TARGET SYSTEM INTO RUN MODE
    .. program code..
    ser_pic16c84(RUN,0);
    .. program code..
```

INC_ADDR increments PIC16CXXX/PIC16F84 data pointer.
BEGIN_PROG programs data at current pointer location.
PARALLEL_MODE puts PIC16CXXX/PIC16F84 into parallel mode.
(not used)
LOAD_DATA_DM loads EEPROM data.
READ_DATA_DM reads EEPROM data.

CONCLUSION

At the risk of repeating myself, I think you're going to find the PIC16CXXX/PIC16F84 microcontrollers some of the easiest to use and most versatile of the Microchip line. That's not to say that other members of the family aren't excellent devices; I only mean to point out the attributes of the PIC16CXXX/PIC16F84s. You will be amazed at how much can be done with these chips. You will like the EEPROM memory that makes software changes so easy. And you are going to love the program-on-the-fly feature that allows remote programming of these devices.

Thus, it's safe to say that you are going to love this programmer. It is simple to construct, simple in nature, and simple to use. All of that adds up to one very pragmatic project that will provide an excellent development tool for your bench. Enjoy!

You will be amazed at how much can be done with these chips.

AN589

```
//************************** FIGURE #2 *******************************
//**                                                               **
//**   SERIAL PROGRAMMING ROUTINE FOR THE PIC16C84 MICROCONTROLLER **
//**                                                               **
//**                   Analog Design Specialists                   **
//**                                                               **
//********************************************************************

//FUNCTION PROTOTYPE: int ser_pic16c84(int cmd, int data)

// cmd: LOAD_CONFIG    -> part configuration bits
//      LOAD_DATA      -> program data, write
//      READ_DATA      -> program data, read
//      INC_ADDR       -> increment to the next address (routine does not auto increment)
//      BEGIN_PROG     -> program a previously loaded program code or data
//      LOAD_DATA_DM   -> load EEPROM data regesters (BEGIN_PROG must follow)
//      READ_DATA_DM   -> read EEPROM data
//
// data: 1) 14 bits of program data or
//       2)  8 bits of EEPROM data (least significant 8 bits of int)

// Additional programmer commands (not part of PIC16C84 programming codes)
//
// cmd: RESET          -> provides 1 ms reset pulse to target system
//      PROGRAM_MODE   -> initializes PIC16C84 for programming
//      RUN            -> disconnects programmer from target system
//
// function returns:1) 14 or 8 bits read back data for read commands
//                  2) zero                for write data commands
//                  3) PIC_PROG_EROR = -1 for programming function errors
//                  4) PROGMR_ERROR  = -2 for programmer function errors


#include <bios.h>

#define LOAD_CONFIG    0
#define LOAD_DATA      2
#define READ_DATA      4
#define INC_ADDR       6
#define BEGIN_PROG     8
#define PARALLEL_MODE 10  // not used
#define LOAD_DATA_DM   3
#define READ_DATA_DM   5
#define MAX_PIC_CMD   63  // division between pic16c84 and programmer commands

#define RESET         64  // external reset command, not needed for programming
#define PROGRAM_MODE  65  // initialize program mode
#define RUN           66  // electrically disconnect programmer

#define PIC_PROG_EROR -1
#define PROGMR_ERROR  -2

#define PTR            0  // use device #0

// parallel port bits
//      d0: data output to part to be programmed
//      d1: programming clock
//      d2: data dirrection, 0= enable tri state buf -> send data to part
//      d3: Vpp control 1= turn on Vpp
//      d4: ~MCLR =0, 1 = reset device with MCLR line
//      d5: clock line tri state control, 0 = enable clock line

int ser_pic16c84(int cmd, int data)            // custom interface for pic16c84
  {
  int i, s_cmd;
```

Figure 3.2.
(Continued on next two pages) **Code for PIC16F84 PC-based programmer.**

AN589

```
if(cmd <=MAX_PIC_CMD)                              // all programming modes
  {
  biosprint(0,8,PTR);                              // set bits 001000, output mode, clock & data low
  s_cmd = cmd;                                     // retain command "cmd"
  for (i=0;i<6;i++)                                // output 6 bits of command
    {
    biosprint(0,(s_cmd&0x1) +2+8,PTR);             // set bits 001010, clock hi
    biosprint(0,(s_cmd&0x1)   +8,PTR);             // set bits 001000, clock low
    s_cmd >>=1;
    }

  if((cmd ==INC_ADDR)||(cmd ==PARALLEL_MODE)       // command only, no data cycle
    return 0;

  else if(cmd ==BEGIN_PROG)                        // program command only, no data cycle
    {
    delay(10);                                     // 10 ms PIC programming time
    return 0;
    }

  else if((cmd ==LOAD_DATA)||(cmd ==LOAD_DATA_DM)||(cmd ==LOAD_CONFIG)) // output 14 bits
    for (i=200;i;i--) ;                            // delay between command & data
    biosprint(0,2+8,PTR);                          // set bits 001010, clock hi; leading bit
    biosprint(0,  8,PTR);                          // set bits 001000, clock low

    for (i=0;i<14;i++)                             // 14 data bits, lsb first
    {
    biosprint(0,(data&0x1) +2+8,PTR);              // set bits 001010,  clock hi
    biosprint(0,(data&0x1)   +8,PTR);              // set bits 001000,  clock low
    data >>=1;
    }
    biosprint(0,2+8,PTR);                          // set bits 001010, clock hi; trailing bit

// ***************  Analog Design Specialists  ******************

    biosprint(0,  8,PTR);                          // set bits 001000, clock low

    return 0;
    }

  else if((cmd ==READ_DATA)||(cmd ==READ_DATA_DM)) //read 14 bits from part, lsb first
    {
    biosprint(0,  4+8,PTR);                        // set bits 001100, clock low, tri state data buffer
    for (i=200;i;i--) ;                            // delay between command & data
    biosprint(0,2+4+8,PTR);                        // set bits 001110, clock hi, leading bit
    biosprint(0,  4+8,PTR);                        // set bits 001100, clock low

    data =0;
    for (i=0;i<14;i++)                             // input 14 bits of data, lsb first
    {
    data >>=1;                                     // shift data for next input bit
    biosprint(0,2+4+8,PTR);                        // set bits 001110,  clock hi
    biosprint(0,  4+8,PTR);                        // set bits 001100,  clock low
    if(!(biosprint(2,0,0)&0x40)) data += 0x2000;   //use printer acknowledge line for input,
                                                   //data lsb first
    }
    biosprint(0,2+4+8,PTR);                        // set bits 001110, clock hi, trailing bit
    biosprint(0,  4+8,PTR);                        // set bits 001100, clock low
    return data;
    }

  else return PIC_PROG_EROR;                       // programmer error

  }
else if(cmd == RESET)                              // reset device
```

AN589

```
      {
   biosprint(0,32+16+4,PTR);              // set bits 110100, MCLR = low
(reset                                    // PIC16C84), programmer not connected
   delay(1);                              // 1ms delay
   biosprint(0,32    +4,PTR);             // set bits 100100, MCLR = high
   return 0;
   }

 else if(cmd ==PROGRAM_MODE)              // enter program mode
   {
   biosprint(0,32+16+4,PTR);              // set bits 110100, Vpp off, MCLR =low
                                          //(reset PIC16C84)
   delay(10);                             //10 ms, allow programming voltage to stabilize

   biosprint(0,8,PTR);                    // set bits 001000, Vpp on , MCLR = 13.5 volts,
                                          // clock & data connected
   delay(10);                             // 10 ms, allow programming voltage to stabilize

   return 0;
   }

 else if(cmd ==RUN)                       // disconnects programmer from device
   {
   biosprint(0,32+4,PTR);                 // set bits 100100
   return 0;
   }
 else return PROGMR_ERROR;                // command error
 }
```

Figure 3.2.
(Continued)

**Figure 3.4.
One of many commercially built programmers available for the PICMICRO® devices. This is microEngineering Labs' EPIC Plus pocket programmer.**

Figure 3.5. microEngineering Labs' EPIC Plus pocket programmer on its plastic stage with a parallel cable used to connect the programmer to a PC.

CHAPTER 4
THE LED FLASHER

INTRODUCTION

There's an old legend that tells the woeful tail of an electronics hobbyist who was experimenting with a new type of microcontroller. Once he understood the basics of the new device and its instruction set, he proceeded to write a terrific program that allowed the microcontroller to monitor the temperature of his personal computer processor. This, he concluded, would provide protection for his system by warning him if the processor was getting too hot.

After considerable debugging, the program was ready to be entered into the microcontroller for a test run. The hobbyist burned the software into the device, plugged it into the hardware, and placed the temperature sensor in the proper place inside his computer. He then fired up the PC and sat back, admiring his new project. Everything was ready for a trial. With the flip of a switch, power was applied to the temperature sensor and, *voila*, the microcontroller, with the dedicated temperature-sensing

This fable reinforces a programming tradition that is almost to the point of being a ritual—that your first endeavor with any new processor is to flash an LED.

code, worked flawlessly! The LCD display came alive with the proper information, and each time the temperature changed, the display accurately indicated that change. Yes, it was a brilliant effort!

The only problem seemed to be that at the instant the hobbyist activated the power switch, a dark and very dense cloud of smoke rose around the chair he was seated in. When the smoke finally cleared, he was no-where to be found. All that was left was a few wisps, from the original thick haze, rising from the chair. He had vanished—never to be seen again, alive or otherwise!

After an exhaustive, but inconclusive, investigation it was determined that the hobbyist had committed the unfor-givable sin of not making his first project, regarding the new microcontroller, a light-emitting diode (LED) flasher. This unacceptable error resulted in his banishment to a as-yet- not-totally-understood place (probably floating in limbo for eternity). There was simply no other rational explanation. Other theories were proposed, but consid-ering the legend of "the curse" associated with not flash-ing an LED first, these were rejected, and the case of this hobbyist's disappearance was closed forever.

The LED flasher is a good initial project because it will ease you into both the hardware and software. It is quick to build and the code is simple to write.

Now, if you believe that story, then I have a bridge you're just going to love! It is in New York City and is named Brooklyn, or something like that, but the name can eas-ily be changed. After all, it will be your bridge. Give me a call at your earliest convenience.

All kidding aside, however, this fable does reinforce a programming tradition that is almost to the point of being a ritual—that your first endeavor with any new processor is to flash an LED. Many an otherwise sane computer experimenter believes that there are severe penalties to be suffered if you depart from that old tradition. So, as not to press our luck (or perhaps anger the computer gods), the first project I'm describing in this book is— you guessed it—a light-emitting diode flasher. What the heck, it can't hurt! Besides, you can never be too careful when it comes to these things.

Actually, this is a good place to start, as this device is very basic in hardware design and code and will provide a sound foundation upon which to build your PICMICRO® microcontroller abode. (Don't you just love all these rhymes and alliterations I come up with? I thought you did. Oh well, at least it helps break up the monotony.)

Where was I? Oh yeah, the LED flasher is a good initial project because it will ease you into both the hardware and software. It is quick to build and the code is simple to write. Hence, hopefully, you won't get bored or frustrated and quit before you have finished your first venture into PICMICRO® microcontrollers.

So, without further fanfare, let's get going. All you will need is a handful of components, your personal computer and your PICMICRO® MCU programmer to put this one on the map. With this effort, you will be on your way to years of excitement and adventure using the Microchip products.

Well, I don't know about *years*. At least *months* of excitement and adventure. Well, maybe years! These devices will get your attention!

THEORY

What we have here is a device that will flash that mystical light-emitting diode at approximately 0.5 hertz. That is, the LED will stay on for half a second, then stay off for half a second. Trust me, this will meet the "flashing an LED" obligation, keeping you, your computer, your family, your dog, and your house safe from "the curse."

HARDWARE

All right, enough with the curse. Let's get down to business. Referring to *Figure 4.1*, the schematic diagram, you will see what qualifies as nothing short of a simple project. With a total of seven components, they don't come much simpler that this one. And there is absolutely nothing even close to being critical about the circuit or the layout. Now, that is my kind of project!

In a very standard configuration, the master clear (MCLR) line, pin 4, is held high by a 4,700-ohm pull-up resistor (R1), and the crystal (XTAL1) is connected across the two oscillator lines (OSC1 and OSC2), pins 15 and 16. Two 22-picofarad capacitors (C1 and C2) are hooked in between the oscillator lines and ground, providing a little extra kick to help the clock get going.

The positive voltage is applied to pin 14 (Vdd), and ground goes to pin 5 (Vss). Last, but maybe most important, the LED connects to input/output (I/O) port RA0, pin 17, through 4,700-ohm "dropping" resistor R2. Naturally, it is this LED that does the all-important flashing! Oooo-Eeee-Oooo!

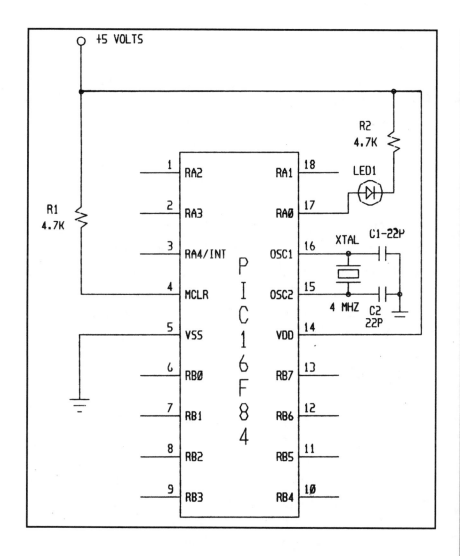

+5 VOLTS

R2
4.7K

R1
4.7K

1	RA2		RA1	18	LED1
2	RA3		RA0	17	
3	RA4/INT		OSC1	16	XTAL C1-22P
4	MCLR	PIC16F84	OSC2	15	4 MHZ C2 22P
5	VSS		VDD	14	
6	RB0		RB7	13	
7	RB1		RB6	12	
8	RB2		RB5	11	
9	RB3		RB4	10	

**Figure 4.1.
Schematic
diagram of the
LED flasher.**

So, there isn't much to the hardware. That, however, is one of the facets of microcontrollers that's so nice. A lot of the work is done by the software, meaning the hardware doesn't have to be complex. I like that, don't you?

As I said, there is nothing critical here. You can construct this gem on a solderless breadboard, a printed circuit

SEMICONDUCTORS:	RESISTORS:
PIC16F84 Microcontroller (1)	4,700 Ohm ¼ Watt (2)
Red Light-Emitting Diode (1)	
4-Megahertz Quartz Crystal (1)	**CAPACITORS:**
	22-Picofarad Ceramic Disk (2)

board (PCB), or use point-to-point wiring on perf-board—whatever strikes your fancy. Probably though, for this circuit, a breadboard will do the trick. For as much as you will love this little LED flasher, it is not likely you will want to keep it forever. Of course, maybe you will, and in that case, use PCB.

SOFTWARE

Next, let's look at the software. Again, nothing complicated or bizarre here. *Figure 4.2* lists the code, and, as can be immediately seen, it isn't hard. It consists of 28 commands that essentially turn on the LED, run a delay routine, turn the LED off, and then "loop" the program so it does all that again. Therefore, the LED goes on and off, which is synonymous with flashing.

Now, I'm not trying to insult your intelligence by being overly simplistic; it's just that I'm writing this as though I were trying to explain it to *myself*! And with me, the simpler the explanation, the better. Since there may well be others like me out there reading this, the approach seems prudent. For the rest of you, please be patient and understanding! It gets better in the coming chapters.

```
; = = = = = = = = = = = = = = = = = = = = = = = = = = = = = = = = = = = = = = = = = = =
; Flash LED
; Flash LED at Approximately Half-Second Cycles, Set Port RA0 High and Low
; - - - - - - - - - - - - - - - - - - - - - - - - - - - - - - - - - - - - - - - - - - -
LIST P=16F84
RADIX HEX
; - - - - - - - - - - - - - - - - - - - - - - - - - - - - - - - - - - - - - - - - - - -
temp            equ         12              ; 16-bit delay
; Flash Routine
                org         0x000
                clrf        porta           ; Clear all bits in porta
start           movlw       0x00            ; Load W with 0x00
                tris        0               ; Copy W tri-state, porta 0
                clrf        porta           ; All lines low
go              bsf         0               ; Turn on LED
                call        delay           ; Run delay
                bcf         0               ; Turn off LED
                call        delay           ; Run delay
                goto        loop            ; Run loop
;  Delay Routine
delay                                       ; Identifies loop routine
                movlw       0               ; Set delay value
                movwf       temp            ; Moves W to temp register
                movlw       128             ; Load W with literal 128
                movwf       temp + 1        ; Decrement temp
                return                      ; End delay
; D_Loop                                    ; Run loop
                decf        temp            ; Decrement low value
                btfsc       status,z        ; Check zero flag
                decf        temp + 1        ; Decrement temp
                movf        temp,W          ; Moves Copy to temp W
                iorwf       temp + 1,W      ; "OR's" W, Results in
                                            ;   temp W
                btfss       status,z        ; Test zero register
                goto        D_loop          ; Goes to loop
return                                      ; Ends loop
end                                         ; Ends program
;========================================================================
```

Figure 4.2.
LED flasher
software.

For a more detailed description of the code, let me go through it line by line. First, anytime a semicolon (;) preceeds a line in the software, that line will be ignored

Anytime a semicolon (;) preceeds a line in the software, that line will be ignored by the assembler. This is what is known as a "delimiter," and it allows you to make notations on the software that will not actually be part of the completed code.

by the assembler. This is what is known as a "delimiter," and it allows you to make notations on the software that will not actually be part of the completed code. Thus, the lines of equal signs (=), the lines that are all dashes (-), and all that stuff about "Flash LED" will be eliminated by the assembler program. They are just window dressing for the code text.

That means that the first lines that are relevant are "LIST P=16F84" and "RADIX HEX". List P=16F84 tells the assembler that a PIC16F84 microcontroller is being used, and the second line specifies the "HEX" format. Next, "temp equ 12" sets up a 16-bit delay, as indicated in the line following the semicolon. "Flash Routine", of course, means nothing to the assembler (it's behind a semicolon, and I think that is sort of like being behind the eight ball).

Now, "org 0x000" sets the starting point, while "clrf porta" clears all the bits in the PIC16F84's A port. The next line, "start movlw 0x00", loads the "W" register with a value of "0x00". This is followed by "tris 0", which copies W tri-state to porta "0" only. Tri-state means an "off" condition. The last "start" command is "clrf porta", and that sets all porta lines low.

Next, we have "go bsf 0", which does an important thing—it turns the LED on. The delay is then called by "call delay," which starts and runs the delay routine. That code will be covered momentarily. Following the delay, "bcf 0" does the second important thing—it turns the LED off. There you go. The flash! To keep the LED off for the requisite half second, however, the delay routine is again called by "call delay". Without this action, the LED would appear to be on all the time. Now, the loop is brought up by the command "goto loop". That, of course, keeps everything running until the power is removed. This is what is termed a "continuous loop," or one that runs over and over again until something, such as shutting off the voltage, stops it.

And that's the basic program! A very simple, but efficient, code. There are, however, two subroutines—"delay" and "loop"—that need to be examined in more detail. Let's start with the delay. The first line in the subroutine, "delay", simply identifies the delay routine to the assembler. Next, "movlw 0" sets the delay value, and "movlw temp" moves the contents of the W register into the "temp", or temporary register.

Now we want to load the "literal" 128 into the W register, and "movlw 128" does the trick. The temp register is then decremented by one with "movwf temp + 1". Finally, the "return" command ends the delay. Once again, nothing complex about this subroutine, but it does the job very well. Here is an excellent example of how efficient the PICMICRO® MCU instruction set really is.

The last part of the code concerns the loop, so let's take a look at that. The line "decf temp" is used to decrement the low temp value, while "btfsc status,z" checks for the "zero" flag. Next, the temporary register is decremented with "decf temp + 1", and the "movf temp,W" moves that value to the temporary W register. The "iorwf temp + 1,W" command then "OR's" the contents of the W register and puts the result in temp W. Now, "btfss status,z" tests the "zero" register, while "goto D_loop" takes the program to the loop. Finally, "return" ends the loop, and "end" ends the program. In this case, however, that will not happen until the power is removed, but "end" is a necessary command in any PICMICRO® MCU program.

And there it is: the very comprehensible code for our "life-saving" LED flasher. Now all that's necessary is to load the code into the PIC16F84, install the PICMICRO® MCU in the hardware, and sit back and wipe the sweat from our brows. If the LED flashes, you're out of the proverbial woods. No dark, dense smoke!

Using the development tools, such as the MPLAB® IDE, to load the code into the PICMICRO® MCUs is covered in detail in the software chapter (Chapter 11). Regarding the hardware, as previously stated, it is a matter of personal preference. In the end, the combination of the programmed PIC16F84 and the completed circuit board will result in a flashing light-emitting diode (see *Figure 4.3*). Trust me, it will happen.

Getting over this hurdle gets you better than halfway to the finish line.

CONCLUSION

Aside from all the fun we have had with the curse, this is a really great first project for anyone interested in PICMICRO® microcontrollers. Understanding the code used for this device goes a long way toward your total comprehension of the 35-command instruction set used by these chips. Hence, getting over this hurdle gets you better than halfway to the finish line (there's another of those amusing alliterations).

Also, the LED flasher helps you understand the operation of the clock, power, and input/output lines found on the

**Figure 4.3.
Here's the final
product—the
"all-important"
LED flasher unit
assembled on a
small solderless
breadboard,
guaranteed to
protect you from
"the curse!"**

PIC16F84 and other PICMICRO® microcontrollers. Again,
this takes you closer to a complete awareness of how
these dear, sweet devices function. I know, I'm getting
sentimental, but I can't help it! It's in my nature! On that
note, enjoy!

CHAPTER 5
A DIGITAL FOUR-CHANNEL VOLTMETER WITH KEYBOARD AND DISPLAY

INTRODUCTION

This project provides an interesting finished product, as well as a highly valuable educational experience. We will be looking at the "full-blown" version, with the analog inputs, LED display, and keyboard, but Microchip application note AN557 (*Figure 5.2*) takes the reader through each step of adding the various features. Additionally, the note covers the software in a step-by-step fashion. So, in getting to our finished product (see *Figure 5.4*), we will explore the basic unit and the additions that produce the final configuration of this handy voltmeter. It is an absorbing process that again illustrates the versatility of the PICMICRO® MCU line.

First, however, let me tell you a little about the goals of this exercise. The voltmeter utilizes one of the midrange, high-speed 8-bit microcontrollers: the PIC16C71, which is a member of the PIC16CXXX line. Some of the worthy features of this chip are the 14-bit instruction word, the

The voltmeter utilizes one of the midrange, high-speed 8-bit microcontrollers: the PIC16C71, which is a member of the PIC16CXXX line.

on-chip 8-bit analog-to-digital (A/D) converter (four-channel), and the interrupt capability.

As for the hardware, the completed meter offers a four-digit LED display for operator feedback, four channels (inputs), and a 16-switch keypad for data entry. Furthermore, the device is completely self-contained and can be fashioned into a relatively small package. When you add all this up, the result is an admirable digital voltmeter that is more than worth the effort.

So, without further ado, let's get our feet wet. This is an enjoyable project—easily and quickly constructed—that effectively portrays the merits of the PIC16C71 microcontroller. Also, if you're in a real hurry, this circuit can be breadboarded in a snap. In that scenario, you will be able to evaluate the device before committing to a more formal version.

HARDWARE THEORY

For this section, refer to *Figure 5.1*. I guess that isn't essential, but I think it will help. Seriously though, one valuable improvement involving the PIC16C71 is the I/O port sink/source parameters. Each input/output (I/O) pin can source up to 20 milliamps and sink 25 milliamps, with total PORTB capability at 100 milliamps sourced and 150 milliamps sinked. That is, at the very least, impressive, and it puts these devices in the range of the digital buffers.

Since we will be using the PORTB to drive the four-digit LED display, those specifications make the PIC16C71

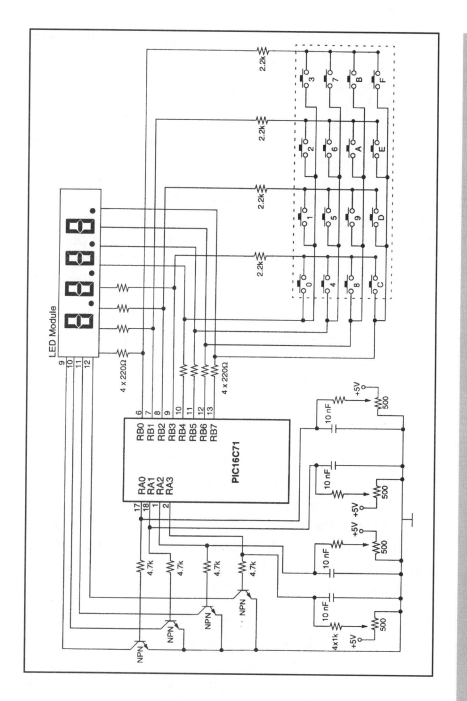

**Figure 5.1.
Schematic
diagram of the
four-channel
digital voltmeter
with keyboard
and display.**

perfect for the task. As seen in Figure 5.1, the eight PIC16C71 PORTB lines are connected to the eight individual segments of the multiplexed LED display (a, b, c, d, e, f, g, and the decimal point). Four external transistors handle digital current sinking of each of the four seven-segment LED displays; however, a transistor array, such as the ULN2003, could also be employed.

PORTA, which manages the base drive of each transistor, is rated at 80 milliamps sink and 50 milliamps source. That, of course, is more than adequate for the purpose they serve. This brings us to a total of 12 I/O lines, which is one less than the maximum of 13 available on this 18-pin microcontroller. Thus, all is well!

Next, let me direct your attention to the keypad. Interfacing this to the circuit, as seen in Figure 5.1, is a simple matter of connecting the "rows" and "columns" to PORTB. The PIC16C71's internal 20-kilohm pull-up resistors on lines RB4 to RB7 are easily enabled/disabled by clearing/setting bit RBPU. The average keystroke will last from about 50 milliseconds to as long as the key is held down. So, to avoid the system missing a keystroke, the keypad is sampled just after the most significant digit (MSD), or every 20 microseconds. Also, the keypad columns are connected to ground through 2.2-kilohm resistors to detect the keystrokes. In the end, all of this serves to integrate the keypad into the design.

Now comes the analog-to-digital (A/D) converter. PORTA lines RA0 to RA3 serve as the four analog channels (inputs) of our voltmeter. Whenever these lines are connected as normal input/output (I/O) ports, they can be

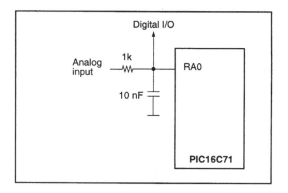

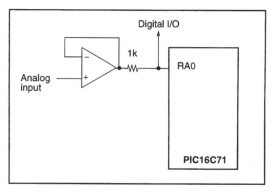

Figure 5.3.
Typical analog-to-digital input buffers.

temporarily put into service as analog inputs. However, a buffer is sometimes necessary to eliminate interference from the analog source. And sometimes it isn't necessary. This often depends on the nature of the analog signal.

Figure 5.3 illustrates a couple of prototypical circuits for this purpose. An "analog input" is the wiper of the potentiometer, as seen in Figure 5.1. The resistance/capacitance (R/C) component is required to smooth the analog voltage. While the R/C does improve the dependability of the analog reading, it also injects a short "time delay", which might be of concern if this configuration were employed in a design with brisker readings.

Okay! Let's review this. The finished product provides four input channels connected through four separate potentiometers. Information to the system can be externally entered through the keypad, and the system results are annunciated on a four-digit, seven-segment LED display. Additionally, the analog inputs are sampled every 20 milliseconds, and that sample rate could be increased up to every five microseconds if necessary. The keyboard sample rate, however, doesn't have to exceed 20 microseconds.

That about covers the hardware aspects of this project. It doesn't take a rocket scientist to see the simplicity of this one. If desired, the whole works will easily go onto a single printed circuit board (PCB), or it could be subdivided into individual modules—whatever the requirements might be. And you will be happy to learn that the software is just about as simplistic. Speaking of which, that's the next topic.

SYSTEM SOFTWARE

As has been stated, this device employs a multiplexed display. If you are not familiar with this type of display, let me explain how it works and why it is used. Multiplexing, here, refers to the process of turning only one seven-segment display unit on at any given time. And, if you turn the four units on and off fast enough, they will all appear to be on all the time.

This is the result of a phenomenon know as persistence of vision. The human brain has the ability to retain an image for a short time after said image has actually

In the case of our voltmeter, persistence of vision comes into play by turning the displays on and off so rapidly that each distinct image doesn't have time to fade from the mind before the next image appears.

disappeared from view. With most people, this retention is normally in the range of two-tenths of a second, or about 200 milliseconds. If a new image is introduced before the last image vanishes, however, the mental perception is a smooth transition from one image to the next. This is why individual motion-picture frames flashing on and off a screen at 24 frames per second appear to be one seamless image.

In the case of our voltmeter, persistence of vision comes into play by turning the displays on and off so rapidly that each distinct image doesn't have time to fade from the mind before the next image appears. In this fashion, the entire display seems to be illuminated at any given moment. This is a very clever scheme, but why go to all this trouble? A very good question, and the following will hopefully provide the answer.

With this design, we are working with light-emitting diodes, and LEDs are, to say the least, power hungry. In the event that you want to make this project portable—you know, battery operated—the current draw of the displays will be a heavy burden on those batteries, unless you multiplex the display. Since only one display is actually on at a time, the current drain is one quarter of what it would be if all four displays were on. Hence, the batteries, at least in theory, will last four times as long.

Also, unlike incandescent lamps, rapidly turning LEDs on and off has little, if any, effect on the longevity of the diodes. With lamps, the constant expanding and contracting of the delicate filaments, as electrical current is applied and removed, will eventually prove to be the lamp's Achilles' heel. Like bending a piece of wire back and forth too many times, the filaments finally break. So, this hazard is not a concern when employing LEDs.

At this point, it might have occurred to you that there is another highly pragmatic reason to employ multiplexing, and that involves the number of port lines needed by the display. With a multiplexed arrangement, the display only requires eight port lines (one for each of the separate segments: a, b, c, d, e, f, g, and the decimal point). If you address each segment individually, however, now we are looking at 32 lines (four a's, four b's, four c's, etc.). Since we only have 13 port lines to work with for this application, driving the individual segments is out of the question.

With that rather long explanation under our belts, let's get to the software. In this circuit, the multiplexing is the product of activating each LED for five microseconds every 20 microseconds. This results in a refresh rate of about 50 hertz, and that is more than adequate for persistence of vision.

The five-microsecond rate is achieved when the microcontroller's prescaler divides the pulse stream from a 4.096-megahertz (4 megahertz for all intents and purposes) clock. The prescaler is set for "divide by 32" and assigned to "timer0" (TMR0). While TMR0 is preloaded with a value of 96, it will increment to FFh, then roll over to 00h, following a "period =". This is seen in the following formula.

$$(256-96) \times (32 \times 4/4096000) = 5 \text{ microseconds}$$

As a result of the TMR0 rollover, flag bit TOIF is set, and an interrupt occurs due to bits TOIE and GIE being enabled. Remember the interrupt capability mentioned above? Now, a simple timer, incrementing every one second, is initiated. With each timer transition, the two eight-bit registers—most significant digit time (MsdT) and least significant digit time (LsdT)—are advanced in a binary decimal code (BCD) format, with the display's least significant digit (LSD) being represented by the lower four bits of LsdT. In turn, the next four bits (high four bits of LsdT) represent the next, or second, significant digit on the display, and this continues until the entire contents of both the LsdT and MsdT registers are employed and all four digits are designated.

So, as the various display digits are activated, the needed information (four-bit BCD format) is transferred from either the MsdT or LsdT register, decoded, and sent to the appropriate seven-segment display. As for the TMR0 interrupt, this has a one-microsecond time and appears exactly every five microseconds. The interrupt service routine can house the entire display-refresh program without any danger of an interrupt within an interrupt.

Since we have added a keypad to this circuit, we will need some code to go along with it. Here is a description of the software required to scan the keypad. Sampling the keys first involves disabling the digital sinks, then configuring PORTB RB4 to RB7 as inputs and RB0 to RB3 as outputs driven high. This enables the pull-up resistors on RB4 to RB7. Next, RB0 to RB3 are consecutively taken low, while RB4 to RB7 are monitored for a keystroke (low level).

The hardware for this project is simple, and the code only consumes memory space of 207 for the program and 13 for data. Hence, there is memory left over in the PIC16C71.

When the program detects a keystroke, a 40-millisecond "debounce" cycle is instituted to prevent the switch from causing multiple entries. As you may be aware, mechanical switches have an annoying habit of "bouncing" when activated. This bouncing is the result of the internal mechanical components settling back into place following the switch action. As they settle, they also tend to make contact more than once. A time delay, or debounce period, alleviates this problem by only allowing the first contact to be detected.

During this debounce period, no other keystrokes will be detected, and keypad sampling will resume only after the 40-millisecond delay has lapsed. This process also serves to minimize false key entries. In review, this code essentially inputs a keystroke and shows the value on the display.

The last of the code to discuss is the analog inputs, and this is super simple. The program monitors the four inputs and saves the values in four consecutive memory locations. The first location is "ADVALUE", which handles channel0. The information is then available when needed by the system. As I said, not much to that.

CONCLUSION

As promised, this is a great project, not only on the surface, but also as a possibility for science fairs, as a teaching aid, or a school project. The hardware is simple, and the code only consumes memory space of 207 for the program and 13 for data. Hence, there is memory left over in the PIC16C71.

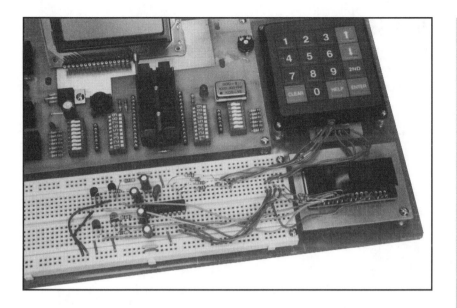

*Figure 5.4.
The final prod-
uct—the four-
channel voltme-
ter, as laid out on
the prototype lab
breadboard. Note
how simple the
hardware is.*

I think you will like working with the PIC16C71 chip. It has many features—some of which we covered in this chapter—that make it an excellent prototype microcontroller. But that certainly doesn't detract from its performance in actual circuit designs. So, as always, have fun with this one!

AN557

APPENDIX D: MPLXAD.ASM

```
MPASM 01.40 Released        MPLXAD.ASM   1-16-1997  16:23:40        PAGE  1


LOC  OBJECT CODE    LINE SOURCE TEXT
     VALUE

                    00001 ;*********************************************************************
                    00002 ;This program demonstrates how to multiplex four 7 segment LED
                    00003 ;digits and a 4X4 keypad along with 4 A/D inputs using a PIC16C71.
                    00004 ;The four digits will first display the decimal a/d value of ch0.
                    00005 ;When keys from 0 - 3 are hit the corresponding channel's a/d value
                    00006 ;is displayed in decimal.
                    00007 ;The LEDs are updated every 20mS, the keypad is scanned at a rate of 20
                    00008 ;mS. All 4 channels are scanned at 20mS rate, so each channel gets
                    00009 ;scanned every 80mS. A faster rate of scanning is possible as required
                    00010 ;by the users application.
                    00011 ;Timer0 is used in internal interrupt mode to generate the
                    00012 ;5 mS.
                    00013 ;
                    00014 ;                                        Stan D'Souza 5/8/93
                    00015 ;
                    00016 ;Corrected error in display routine.
                    00017 ;                                        Stan D'Souza 2/27/94
                    00018 ;
                    00019 ;        Program:         MPLXAD.ASM
                    00020 ;        Revision Date:
                    00021 ;                          1-15-97      Compatibility with MPASMWIN 1.40
                    00022 ;
                    00023 ;*********************************************************************
                    00024        LIST P=16C71
                    00025        ERRORLEVEL  -302
                    00026 ;
                    00027        include      <p16c71.inc>
                    00001        LIST
                    00002 ; P16C71.INC Standard Header File, Ver. 1.00 Microchip Technology, Inc.
                    00142        LIST
                    00028 ;
0000000C            00029 TempC    equ      0x0c           ;temp general purpose regs
0000000D            00030 TempD    equ      0x0d
0000000E            00031 TempE    equ      0x0e
00000020            00032 PABuf    equ      0x20
00000021            00033 PBBuf    equ      0x21
0000000F            00034 Count    equ      0x0f           ;count
00000010            00035 MsdTime  equ      0x10           ;most significant Timer
00000011            00036 LsdTime  equ      0x11           ;Least significant Timer
                    00037 ;
00000012            00038 Flag     equ      0x12           ;general purpose flag reg
                    00039 #define keyhit    Flag,0         ;bit 0 --> key-press on
                    00040 #define DebnceOn  Flag,1         ;bit 1 -> debounce on
                    00041 #define noentry   Flag,2         ;no key entry = 0
                    00042 #define ServKey   Flag,3         ;bit 3 --> service key
                    00043 #define ADOver    Flag,4         ;bit 4 --> a/d conv. over
                    00044 ;
00000013            00045 Debnce   equ      0x13           ;debounce counter
00000014            00046 NewKey   equ      0x14
00000015            00047 DisplayCh equ     0x15           ;channel to be displayed
                    00048 ;
00000016            00049 ADTABLE equ       0x16           ;4 locations are reserved here
                    00050                                  ;from 0x16 to 0x19
                    00051 ;
```

Figure 5.2.
(Continued on next seven pages)
Code for the digital four-channel voltmeter.

AN557

```
0000002F          00052 WBuffer  equ     0x2f
0000002E          00053 StatBuffer equ   0x2e
00000001          00054 OptionReg equ    1
00000002          00055 PCL      equ     2
                  00056 ;
                  00057 ;
                  00058 push    macro
                  00059         movwf   WBuffer        ;save w reg in Buffer
                  00060         swapf   WBuffer, F     ;swap it
                  00061         swapf   STATUS,W       ;get status
                  00062         movwf   StatBuffer     ;save it
                  00063         endm
                  00064 ;
                  00065 pop     macro
                  00066         swapf   StatBuffer,W   ;restore status
                  00067         movwf   STATUS         ;       /
                  00068         swapf   WBuffer,W      ;restore W reg
                  00069         endm
                  00070 ;
0000              00071         org     0
0000 280D         00072         goto    Start          ;skip over interrupt vector
                  00073 ;
0004              00074         org     4
                  00075 ;It is always a good practice to save and restore the w reg,
                  00076 ;and the status reg during a interrupt.
                  00077         push
0004 00AF      M          movwf   WBuffer        ;save w reg in Buffer
0005 0EAF      M          swapf   WBuffer, F     ;swap it
0006 0E03      M          swapf   STATUS,W       ;get status
0007 00AE      M          movwf   StatBuffer     ;save it
0008 2052         00078         call    ServiceInterrupts
                  00079         pop
0009 0E2E      M          swapf   StatBuffer,W   ;restore status
000A 0083      M          movwf   STATUS         ;       /
000B 0E2F      M          swapf   WBuffer,W      ;restore W reg
000C 0009         00080         retfie
                  00081 ;
000D              00082 Start
000D 203B         00083         call    InitPorts
000E 20EE         00084         call    InitAd
000F 2045         00085         call    InitTimers
0010              00086 loop
0010 1992         00087         btfsc   ServKey        ;key service pending
0011 2015         00088         call    ServiceKey     ;yes then service
0012 1A12         00089         btfsc   ADOver         ;a/d pending?
0013 2028         00090         call    ServiceAD      ;yes the service a/d
0014 2810         00091         goto    loop
                  00092 ;
                  00093 ;ServiceKey, does the software service for a keyhit. After a key
                  00094 ;service, the ServKey flag is reset, to denote a completed operation.
0015              00095 ServiceKey
0015 1192         00096         bcf     ServKey        ;reset service flag
0016 0814         00097         movf    NewKey,W       ;get key value
0017 3C03         00098         sublw   3              ;key > 3?
0018 1C03         00099         btfss   STATUS,C       ;no then skip
0019 0008         00100         return                 ;else ignore key
001A 0814         00101         movf    NewKey,W
001B 0095         00102         movwf   DisplayCh      ;load new channel
                  00103 ;
001C              00104 LoadAD
001C 3016         00105         movlw   ADTABLE        ;get top of table
001D 0715         00106         addwf   DisplayCh,W    ;add offset
001E 0084         00107         movwf   FSR            ;init FSR
001F 0800         00108         movf    0,W            ;get a/d value
0020 00A1         00109         movwf   L_byte
0021 01A0         00110         clrf    H_byte
```

AN557

```
0022 2106    00111         call    B2_BCD
0023 0824    00112         movf    R2,W          ;get LSd
0024 0091    00113         movwf   LsdTime       ;save in LSD
0025 0823    00114         movf    R1,W          ;get Msd
0026 0090    00115         movwf   MsdTime       ;save in Msd
0027 0008    00116         return
             00117 ;
             00118 ;This routine essentially loads the ADRES value in the table location
             00119 ;determined by the channel offset. If channel 0 then ADRES is saved
             00120 ;in location ADTABLE. If channel 1 then ADRES is saved at ADTABLE + 1.
             00121 ;and so on.
0028         00122 ServiceAD
0028 0808    00123         movf    ADCON0,W      ;get adcon0
0029 008C    00124         movwf   TempC         ;save in temp
002A 3008    00125         movlw   B'00001000'   ;select next channel
002B 0708    00126         addwf   ADCON0,W      ;          /
002C 1A88    00127         btfsc   ADCON0,5      ;if <= ch3
002D 30C1    00128         movlw   B'11000001'   ;select ch0
002E 0088    00129         movwf   ADCON0
             00130 ;now load adres in the table
002F 3016    00131         movlw   ADTABLE
0030 0084    00132         movwf   FSR           ;load FSR with top
0031 0C8C    00133         rrf     TempC, F
0032 0C8C    00134         rrf     TempC, F
0033 0C0C    00135         rrf     TempC,W       ;get in w reg
0034 3903    00136         andlw   3             ;mask off all but last 2
0035 0784    00137         addwf   FSR, F        ;add offset to table
0036 0809    00138         movf    ADRES,W       ;get a/d value
0037 0080    00139         movwf   0             ;load indirectly
0038 1212    00140         bcf     ADOver        ;clear flag
0039 201C    00141         call    LoadAD        ;load a/d value in display reg.
003A 0008    00142         return
             00143
             00144
             00145
             00146 ;
003B         00147 InitPorts
003B 1683    00148         bsf     STATUS,RP0    ;select Bank1
003C 3003    00149         movlw   3             ;make RA0-3 digital I/O
003D 0088    00150         movwf   ADCON1        ;          /
003E 0185    00151         clrf    TRISA         ;make RA0-4 outputs
003F 0186    00152         clrf    TRISB         ;make RB0-7 outputs
0040 1283    00153         bcf     STATUS,RP0    ;select Bank0
0041 0185    00154         clrf    PORTA         ;make all outputs low
0042 0186    00155         clrf    PORTB         ;          /
0043 1585    00156         bsf     PORTA,3       ;enable MSB digit sink
0044 0008    00157         return
             00158 ;
             00159 ;
             00160 ;The clock speed is 4.096Mhz. Dividing internal clk. by a 32 prescaler,
             00161 ;the TMR0 will be incremented every 31.25uS. If TMR0 is preloaded
             00162 ;with 96, it will take (256-96)*31.25uS to overflow i.e. 5mS. So the
             00163 ;end result is that we get a TMR0 interrupt every 5mS.
0045         00164 InitTimers
0045 0190    00165         clrf    MsdTime       ;clr timers
0046 0191    00166         clrf    LsdTime       ;          /
0047 0195    00167         clrf    DisplayCh     ;show channel 0
0048 0192    00168         clrf    Flag          ;clr all flags
0049 1683    00169         bsf     STATUS,RP0    ;select Bank1
004A 3084    00170         movlw   B'10000100'   ;assign ps to TMR0
004B 0081    00171         movwf   OptionReg     ;ps = 32
004C 1283    00172         bcf     STATUS,RP0    ;select Bank0
004D 3020    00173         movlw   B'00100000'   ;enable TMR0 interrupt
004E 008B    00174         movwf   INTCON        ;
004F 3060    00175         movlw   .96           ;preload TMR0
0050 0081    00176         movwf   TMR0          ;start counter
```

AN557

```
0051 0009       00177          retfie
                00178 ;
0052            00179 ServiceInterrupts
0052 190B       00180          btfsc    INTCON,T0IF      ;TMR0 interrupt?
0053 2857       00181          goto     ServiceTMR0      ;yes then service
0054 018B       00182          clrf     INTCON           ;else clr all int
0055 168B       00183          bsf      INTCON,T0IE
0056 0008       00184          return
                00185 ;
0057            00186 ServiceTMR0
0057 3060       00187          movlw    .96              ;initialize TMR0
0058 0081       00188          movwf    TMR0
0059 110B       00189          bcf      INTCON,T0IF      ;clr int flag
005A 1805       00190          btfsc    PORTA,0          ;scan keys every 20 mS
005B 2060       00191          call     ScanKeys         ;when digit 1 is on
005C 1985       00192          btfsc    PORTA,3          ;scan a/d every 20mS
005D 20F1       00193          call     SampleAd         ;when digit 4 is on
005E 20BF       00194          call     UpdateDisplay    ;update display
005F 0008       00195          return
                00196 ;
                00197 ;
                00198 ;ScanKeys, scans the 4x4 keypad matrix and returns a key value in
                00199 ;NewKey (0 - F) if a key is pressed, if not it clears the keyhit flag.
                00200 ;Debounce for a given keyhit is also taken care of.
                00201 ;The rate of key scan is 20mS with a 4.096Mhz clock.
0060            00202 ScanKeys
0060 1C92       00203          btfss    DebnceOn         ;debounce on?
0061 2866       00204          goto     Scan1            ;no then scan keypad
0062 0B93       00205          decfsz   Debnce, F        ;else dec debounce time
0063 0008       00206          return                    ;not over then return
0064 1092       00207          bcf      DebnceOn         ;over, clr debounce flag
0065 0008       00208          return                    ;and return
0066            00209 Scan1
0066 20A8       00210          call     SavePorts        ;save port values
0067 30EF       00211          movlw    B'11101111'      ;init TempD
0068 008D       00212          movwf    TempD
0069            00213 ScanNext
0069 0806       00214          movf     PORTB,W          ;read to init port
006A 100B       00215          bcf      INTCON,RBIF      ;clr flag
006B 0C8D       00216          rrf      TempD, F         ;get correct column
006C 1C03       00217          btfss    STATUS,C         ;if carry set?
006D 2880       00218          goto     NoKey            ;no then end
006E 080D       00219          movf     TempD,W          ;else output
006F 0086       00220          movwf    PORTB            ;low column scan line
0070 0000       00221          nop
0071 1C0B       00222          btfss    INTCON,RBIF      ;flag set?
0072 2869       00223          goto     ScanNext         ;no then next
0073 1812       00224          btfsc    keyhit           ;last key released?
0074 287E       00225          goto     SKreturn         ;no then exit
0075 1412       00226          bsf      keyhit           ;set new key hit
0076 0E06       00227          swapf    PORTB,W          ;read port
0077 008E       00228          movwf    TempE            ;save in TempE
0078 2082       00229          call     GetKeyValue      ;get key value 0 - F
0079 0094       00230          movwf    NewKey           ;save as New key
007A 1592       00231          bsf      ServKey          ;set service flag
007B 1492       00232          bsf      DebnceOn         ;set flag
007C 3004       00233          movlw    4
007D 0093       00234          movwf    Debnce           ;load debounce time
007E            00235 SKreturn
007E 20B5       00236          call     RestorePorts     ;restore ports
007F 0008       00237          return
                00238 ;
0080            00239 NoKey
0080 1012       00240          bcf      keyhit           ;clr flag
0081 287E       00241          goto     SKreturn
                00242 ;
```

AN557

```
            00243 ;GetKeyValue gets the key as per the following layout
            00244 ;
            00245 ;                    Col1    Col2    Col3    Col4
            00246 ;                    (RB3)   (RB2)   (RB1)   (RB0)
            00247 ;
            00248 ;Row1(RB4)             0       1       2       3
            00249 ;
            00250 ;Row2(RB5)             4       5       6       7
            00251 ;
            00252 ;Row3(RB6)             8       9       A       B
            00253 ;
            00254 ;Row4(RB7)             C       D       E       F
            00255 ;
0082        00256 GetKeyValue
0082 018C   00257         clrf    TempC
0083 1D8D   00258         btfss   TempD,3         ;first column
0084 288C   00259         goto    RowValEnd
0085 0A8C   00260         incf    TempC, F
0086 1D0D   00261         btfss   TempD,2         ;second col.
0087 288C   00262         goto    RowValEnd
0088 0A8C   00263         incf    TempC, F
0089 1C8D   00264         btfss   TempD,1         ;3rd col.
008A 288C   00265         goto    RowValEnd
008B 0A8C   00266         incf    TempC, F        ;last col.
008C        00267 RowValEnd
008C 1C0E   00268         btfss   TempE,0         ;top row?
008D 2896   00269         goto    GetValCom       ;yes then get 0,1,2&3
008E 1C8E   00270         btfss   TempE,1         ;2nd row?
008F 2895   00271         goto    Get4567         ;yes the get 4,5,6&7
0090 1D0E   00272         btfss   TempE,2         ;3rd row?
0091 2893   00273         goto    Get89ab         ;yes then get 8,9,a&b
0092        00274 Getcdef
0092 150C   00275         bsf     TempC,2         ;set msb bits
0093        00276 Get89ab
0093 158C   00277         bsf     TempC,3         ;          /
0094 2896   00278         goto    GetValCom       ;do common part
0095        00279 Get4567
0095 150C   00280         bsf     TempC,2
0096        00281 GetValCom
0096 080C   00282         movf    TempC,W
0097 0782   00283         addwf   PCL, F
0098 3400   00284         retlw   0
0099 3401   00285         retlw   1
009A 3402   00286         retlw   2
009B 3403   00287         retlw   3
009C 3404   00288         retlw   4
009D 3405   00289         retlw   5
009E 3406   00290         retlw   6
009F 3407   00291         retlw   7
00A0 3408   00292         retlw   8
00A1 3409   00293         retlw   9
00A2 340A   00294         retlw   0a
00A3 340B   00295         retlw   0b
00A4 340C   00296         retlw   0c
00A5 340D   00297         retlw   0d
00A6 340E   00298         retlw   0e
00A7 340F   00299         retlw   0f
            00300 ;
            00301 ;SavePorts, saves the porta and portb condition during a key scan
            00302 ;operation.
00A8        00303 SavePorts
00A8 0805   00304         movf    PORTA,W         ;Get sink value
00A9 00A0   00305         movwf   PABuf           ;save in buffer
00AA 0185   00306         clrf    PORTA           ;disable all sinks
00AB 0806   00307         movf    PORTB,W         ;get port b
00AC 00A1   00308         movwf   PBBuf           ;save in buffer
```

AN557

```
00AD 30FF      00309          movlw    0xff             ;make all high
00AE 0086      00310          movwf    PORTB            ;on port b
00AF 1683      00311          bsf      STATUS,RP0       ;select Bank1
00B0 1381      00312          bcf      OptionReg,7      ;enable pull ups
00B1 30F0      00313          movlw    B'11110000'      ;port b hi nibble inputs
00B2 0086      00314          movwf    TRISB            ;lo nibble outputs
00B3 1283      00315          bcf      STATUS,RP0       ;Bank0
00B4 0008      00316          return
               00317 ;
               00318 ;RestorePorts, restores the condition of porta and portb after a
               00319 ;key scan operation.
00B5           00320 RestorePorts
00B5 0821      00321          movf     PBBuf,W          ;get port b
00B6 0086      00322          movwf    PORTB
00B7 0820      00323          movf     PABuf,W          ;get port a value
00B8 0085      00324          movwf    PORTA
00B9 1683      00325          bsf      STATUS,RP0       ;select Bank1
00BA 1781      00326          bsf      OptionReg,7      ;disable pull ups
00BB 0185      00327          clrf     TRISA            ;make port a outputs
00BC 0186      00328          clrf     TRISB            ;as well as PORTB
00BD 1283      00329          bcf      STATUS,RP0       ;Bank0
00BE 0008      00330          return
               00331 ;
               00332 ;
00BF           00333 UpdateDisplay
00BF 0805      00334          movf     PORTA,W          ;present sink value in w
00C0 0185      00335          clrf     PORTA            ;disable all digits sinks
00C1 390F      00336          andlw    0x0f
00C2 008C      00337          movwf    TempC            ;save sink value in tempC
00C3 160C      00338          bsf      TempC,4          ;preset for lsd sink
00C4 0C8C      00339          rrf      TempC, F         ;determine next sink value
00C5 1C03      00340          btfss    STATUS,C         ;c=1?
00C6 118C      00341          bcf      TempC,3          ;no then reset LSD sink
00C7 180C      00342          btfsc    TempC,0          ;else see if Msd
00C8 28D6      00343          goto     UpdateMsd        ;yes then do Msd
00C9 188C      00344          btfsc    TempC,1          ;see if 3rdLsd
00CA 28D3      00345          goto     Update3rdLsd     ;yes then do 3rd Lsd
00CB 190C      00346          btfsc    TempC,2          ;see if 2nd Lsd
00CC 28D0      00347          goto     Update2ndLsd     ;yes then do 2nd lsd
00CD           00348 UpdateLsd
00CD 0811      00349          movf     LsdTime,W        ;get Lsd in w
00CE 390F      00350          andlw    0x0f             ;          /
00CF 28D8      00351          goto     DisplayOut
00D0           00352 Update2ndLsd
00D0 0E11      00353          swapf    LsdTime,W        ;get 2nd Lsd in w
00D1 390F      00354          andlw    0x0f             ;mask rest
00D2 28D8      00355          goto     DisplayOut       ;enable display
00D3           00356 Update3rdLsd
00D3 0810      00357          movf     MsdTime,W        ;get 3rd Lsd in w
00D4 390F      00358          andlw    0x0f             ;mask low nibble
00D5 28D8      00359          goto     DisplayOut       ;enable display
00D6           00360 UpdateMsd
00D6 0E10      00361          swapf    MsdTime,W        ;get Msd in w
00D7 390F      00362          andlw    0x0f             ;mask rest
00D8           00363 DisplayOut
00D8 20DD      00364          call     LedTable         ;get digit output
00D9 0086      00365          movwf    PORTB            ;drive leds
00DA 080C      00366          movf     TempC,W          ;get sink value in w
00DB 0085      00367          movwf    PORTA
00DC 0008      00368          return
               00369 ;
               00370 ;
00DD           00371 LedTable
00DD 0782      00372          addwf    PCL, F           ;add to PC low
00DE 343F      00373          retlw    B'00111111'      ;led drive for 0
00DF 3406      00374          retlw    B'00000110'      ;led drive for 1
```

AN557

```
00E0 345B    00375         retlw    B'01011011'    ;led drive for 2
00E1 344F    00376         retlw    B'01001111'    ;led drive for 3
00E2 3466    00377         retlw    B'01100110'    ;led drive for 4
00E3 346D    00378         retlw    B'01101101'    ;led drive for 5
00E4 347D    00379         retlw    B'01111101'    ;led drive for 6
00E5 3407    00380         retlw    B'00000111'    ;led drive for 7
00E6 347F    00381         retlw    B'01111111'    ;led drive for 8
00E7 3467    00382         retlw    B'01100111'    ;led drive for 9
00E8 3477    00383         retlw    B'01110111'    ;led drive for A
00E9 347C    00384         retlw    B'01111100'    ;led drive for b
00EA 3439    00385         retlw    B'00111001'    ;led drive for C
00EB 345E    00386         retlw    B'01011110'    ;led drive for d
00EC 3479    00387         retlw    B'01111001'    ;led drive for E
00ED 3471    00388         retlw    B'01110001'    ;led drive for F
             00389
             00390 ;
             00391 ;
00EE         00392 InitAd
00EE 30C0    00393         movlw    B'11000000'    ;internal rc for tad
00EF 0088    00394         movwf    ADCON0         ;          /
             00395         ;note that adcon1 is set in InitPorts
00F0 0008    00396         return
             00397 ;
00F1         00398 SampleAd
00F1 20A8    00399         call     SavePorts
00F2 20F8    00400         call     DoAd           ;do a ad conversion
00F3         00401 AdDone
00F3 1908    00402         btfsc    ADCON0,GO      ;ad done?
00F4 28F3    00403         goto     AdDone         ;no then loop
00F5 1612    00404         bsf      ADOver         ;set a/d over flag
00F6 20B5    00405         call     RestorePorts   ;restore ports
00F7 0008    00406         return
             00407 ;
             00408 ;
00F8         00409 DoAd
00F8 0186    00410         clrf     PORTB          ;turn off leds
00F9 1683    00411         bsf      STATUS,RP0     ;select Bank1
00FA 300F    00412         movlw    0x0f           ;make port a hi-Z
00FB 0085    00413         movwf    TRISA          ;          /
00FC 1283    00414         bcf      STATUS,RP0     ;select Bank0
00FD 1408    00415         bsf      ADCON0,ADON    ;start a/d
00FE 307D    00416         movlw    .125
00FF 2102    00417         call     Wait
0100 1508    00418         bsf      ADCON0,GO      ;start conversion
0101 0008    00419         return
             00420 ;
             00421 ;
0102         00422 Wait
0102 008C    00423         movwf    TempC          ;store in temp
0103         00424 Next
0103 0B8C    00425         decfsz   TempC, F
0104 2903    00426         goto     Next
0105 0008    00427         return
             00428
             00429 ;
             00430 ;
00000026     00431 count  equ     26
00000027     00432 temp   equ     27
             00433 ;
00000020     00434 H_byte equ     20
00000021     00435 L_byte equ     21
00000022     00436 R0     equ     22              ; RAM Assignments
00000023     00437 R1     equ     23
00000024     00438 R2     equ     24
             00439 ;
             00440 ;
```

AN557

```
0106 1003      00441 B2_BCD  bcf      STATUS,0      ; clear the carry bit
0107 3010      00442         movlw    .16
0108 00A6      00443         movwf    count
0109 01A2      00444         clrf     R0
010A 01A3      00445         clrf     R1
010B 01A4      00446         clrf     R2
010C 0DA1      00447 loop16  rlf      L_byte, F
010D 0DA0      00448         rlf      H_byte, F
010E 0DA4      00449         rlf      R2, F .
010F 0DA3      00450         rlf      R1, F
0110 0DA2      00451         rlf      R0, F
               00452 ;
0111 0BA6      00453         decfsz   count, F
0112 2914      00454         goto     adjDEC
0113 3400      00455         RETLW    0
               00456 ;
0114 3024      00457 adjDEC  movlw    R2
0115 0084      00458         movwf    FSR
0116 211E      00459         call     adjBCD
               00460 ;
0117 3023      00461         movlw    R1
0118 0084      00462         movwf    FSR
0119 211E      00463         call     adjBCD
               00464 ;
011A 3022      00465         movlw    R0
011B 0084      00466         movwf    FSR
011C 211E      00467         call     adjBCD
               00468 ;
011D 290C      00469         goto     loop16
               00470 ;
011E 3003      00471 adjBCD  movlw    3
011F 0700      00472         addwf    0,W
0120 00A7      00473         movwf    temp
0121 19A7      00474         btfsc    temp,3        ; test if result > 7
0122 0080      00475         movwf    0
0123 3030      00476         movlw    30
0124 0700      00477         addwf    0,W
0125 00A7      00478         movwf    temp
0126 1BA7      00479         btfsc    temp,7        ; test if result > 7
0127 0080      00480         movwf    0             ; save as MSD
0128 3400      00481         RETLW    0
               00482 ;
               00483 ;
               00484 ;
               00485 ;
               00486
               00487         end
```

```
MEMORY USAGE MAP ('X' = Used,  '-' = Unused)

0000 : X---XXXXXXXXXXXX XXXXXXXXXXXXXXXX XXXXXXXXXXXXXXXX XXXXXXXXXXXXXXXX
0040 : XXXXXXXXXXXXXXXX XXXXXXXXXXXXXXXX XXXXXXXXXXXXXXXX XXXXXXXXXXXXXXXX
0080 : XXXXXXXXXXXXXXXX XXXXXXXXXXXXXXXX XXXXXXXXXXXXXXXX XXXXXXXXXXXXXXXX
00C0 : XXXXXXXXXXXXXXXX XXXXXXXXXXXXXXXX XXXXXXXXXXXXXXXX XXXXXXXXXXXXXXXX
0100 : XXXXXXXXXXXXXXXX XXXXXXXXXXXXXXXX XXXXXXXXX------- ----------------

All other memory blocks unused.

Program Memory Words Used:    294
Program Memory Words Free:    730


Errors   :    0
Warnings :    0 reported,     0 suppressed
Messages :    0 reported,     7 suppressed
```

CHAPTER 6
IMPLEMENTING READ TABLES IN PIC16CXXX PROJECTS

INTRODUCTION

Here's a subject that will come in very handy as you delve into PICMICRO® microcontrollers. This material comes from Microchip's application note AN556 and will provide the knowledge of how to access and use "lookup tables" when writing software for your microcontroller projects.

Lookup tables are an extremely handy way to include information in a program that can easily be referenced. For example, when designing a device to read the Dual Tone Multifrequency (DTMF) tone pairs used to dial the touchtone-style telephones we have all become so used to, a read table can be employed for that identification. The hardware will covert the incoming tones to a value that matches one of the listings in the read table, and that will allow the microcontroller to send the appropriate code to a display. Naturally, the display will indicate which number or symbol was dialed.

In this fashion, any device that produces DTMF tones can be monitored in terms of what tone pairs are being issued. The advantages of such a system regarding telephones is more than obvious, but there are a number of other applications for DTMF tones. And it is often useful to monitor them. Some examples include remote-control devices for radio repeater stations, control systems that are accessed by telephone, and monitoring DTMF information sent by wireless means.

Hence, the "DTMF Decoder" example would be one way a read or lookup table can be handily employed in software. I could name a number of additional examples, but I think you get the idea. So, let's move on to a closer look at how all this is done. In the end, I think you will come to appreciate this approach when writing PICMICRO® MCU programs.

IMPLEMENTATION

The Microchip notes provide a series of examples that illustrate how one goes about including and accessing the tables. They also cover some of the pitfalls that must be avoided for the procedure to work. Basically, a table read must be performed. Each table consists of various "retlw K" instructions, and the information is assigned to the "literal K". An "addwf pcl" instruction first determines the table offset, and the program then branches to the needed retwl K instruction.

Example 1 (*Figure 6.2*), which follows, illustrates the fundamentals of this process; however, it does not tell the whole story. When dealing with PIC16CXXX chips, there is some forethought necessary for this to work.

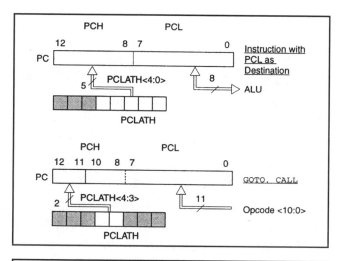

*Figure 6.1.
Table for pro-
gram-counter
(PC) loading
under different
situations.*

```
        .
        .
        .
    movlw   offset  ;load offset in w reg
    call    Table
        .
        .
        .
Table:
    addwf   pcl     ;add offset to pc to
                    ;generate a computed goto
    retlw   'A'     ;return the ASCII char A
    retlw   'B'     ;return the ASCII char B
    retlw   'C'     ;return the ASCII char C
        .
        .
        .
```

*Figure 6.2.
Example 1:
program-counter
loading under
different situa-
tions.*

PROGRAM COUNTER LOADING

The PIC16CXXX program counter (PC) is 13 bits wide, with the lower 8 bits (PCL) in RAM at the 02h locations. These are readable and writable directly, while the higher 5 bits are not. They can only be written to with the PCLATH register. *Figure 6.1* shows how this is done. The PCLATH register is a unique read/write (R/W) register in which only five of the bits are used. The remaining bits take on a "0" value.

With that said, let's look at the mechanics of this in more detail. Two instructions virtually essential to the process are "call" and "goto", although goto is used a little less. If you execute either of these instructions, the lower 11 bits will be provided by the instruction "opcode". The high two bits come from bits 3 and 4 of the PCLATH register. And it is highly advised to preload PCLATH with the high byte of the routine location before running the routine. Example 2 (*Figure 6.3*) explains how.

Example 3 (*Figure 6.4*) illustrates how to preload the PCLATH register with the high byte of the table address. Doing this will help keep the program from branching erratically during a table read.

The problem here is that when the PCL is the target of an instruction, the PCH will be loaded with the PCLATH's five low bits. If, however, the "call" is made to a page other than the address of the table, the "goto" will go to the wrong page. And that will cause trouble. To avoid this, use the scheme in Example 3.

Another problem associated with a table read involves page boundaries. An "addwf pcl" instruction can only handle 8 bits, and the remaining code will end up in an unintended portion of the overall code. Example 4 (*Figure 6.5*) illustrates how this can happen.

Consequently, you have to either take notice as to where in a page the table resides, or add the rollover to the PCLATH before the computed "goto" instruction.

```
            .
            .
    movlw   HIGH Table   ;load high 8-bit
                         ;address of Table
    movwf   PCLATH       ;into PCLATH
    call    Routine      ;execute Call
                         ;instruction
            .
            .
```

> **Note:** If the program memory size is less than 2K-words, then the above precaution is not necessary.

Figure 6.3. Example 2: program-counter loading under different situations.

Example 5 (*Figure 6.6*) shows how to handle both the page boundary and the table location. This requires a 13-bit computed "goto" operation, and Example 5 illustrates the procedure. Hence, this will become very valuable code when utilizing read tables.

```
            .
    org     0x80     ;code location in page 0
    movlw   offset   ;load offset in w reg
    call    Table
            .
            .
    org     0x0320   ;Table located in page 3
Table:
    addwf   pcl      ;add offset to pc to
                     ;generate a computed goto
    retlw   'A'      ;return the ASCII char A
    retlw   'B'      ;return the ASCII char B
    retlw   'C'      ;return the ASCII char C
            .
            .
            .
```

Figure 6.4. Example 3: program-counter loading under different situations.

Figure 6.5.
Example 4:
program-counter
loading under
different situa-
tions.

```
         .
         org     0x80         ;code location in
                              ; page 0
         movlw   HIGH Table   ;load PCLATH with hi
                              ; address
         movwf   PCLATH       ;       /
         movlw   offset       ;load offset in w reg
         call    Table
         .

         .
         org     0x02ff       ;Table located end of
                              ; page 2
Table:
         addwf   pcl          ;value in pc will not
                              ; roll over to page 3
         retlw   'A'          ;return the ASCII
                              ; char A
         retlw   'B'          ;return the ASCII
                              ; char B
         retlw   'C'          ;return the ASCII
                              ; char C
         .

         .

         .
```

Our last example, Example 6 (*Figure 6.7*), addresses how to cope with the difference in the PICMICRO® microcontrollers. Thus far, we have talked about the PIC16CXXX line of devices, but read tables can also be used with the PIC16C5X and PIC17C42 chips. The following will explain how. PIC16C5X devices *do not* have PCH or PCLATH registers, and, needless to say, that will cause some difficulty. In the PIC16C5X family, the table has to be in the upper half of a 512-word page, a restriction that doesn't apply to the PIC16CXXX chips.

For the PIC17C42 microcontroller, the PCLATH register is loaded with the PCH during "goto" or "call". Furthermore, a computed "goto" will not manage a page-boundary crossing.

```
            .
    movlw   LOW Table   ;get low 8 bits of
                        ; address
    addwf   offset      ;do an 8-bit add
                        ; operation
    movlw   HIGH Table  ;get high 5 bits of
                        ; address
    btfsc   status,c    ;page crossed?
    addlw   1           ;yes then increment
                        ; high address
    movwf   PCLATH      ;load high address in
                        ; latch
    movf    offset,w    ;load computed offset
                        ; in w reg
    call    Table
            .
            .
            .
Table:
    movwf   pcl         ;load computed offset
                        ; in PCL
    retlw   'A'         ;return the ASCII
                        ; char A
    retlw   'B'         ;return the ASCII
                        ; char B
    retlw   'C'         ;return the ASCII
                        ; char C
            .
            .
            .
```

Figure 6.6. Example 5: program-counter loading under different situations.

To handle this, a 16-bit computed jump address needs to be calculated before you execute the table read. Here is where Example 6 comes into play. It will allow location of the table anywhere in memory. However, Microchip is quick to caution that the very code-efficient "tablrd/tlrd" instruction should be used for tables that cross page boundaries or are very large.

CONCLUSION

So, there you have it—some very handy code that will allow you to use "read" or "lookup" tables in your programs.

```
          .
          movlw   LOW Table     ;get low 8 bits of
                                ; address
          addwf   offset        ;do an 8-bit add
                                ; operation
          movlw   HIGH Table    ;get high 8 bits of
                                ; address
          btfsc   statist,c     ;page crossed?
          addlw   1             ;yes then increment
                                ; high address
          movwf   PchBuffer,w   ;load in temporary
                                ;location
          call    Table
          .
          .
       Table:
          movf    PchBuffer,w   ;get high offset
          movwf   PCLATH        ;load in latch
          movf    offset,w      ;get low offset
          movwf   pcl           ;load computed
                                ; offset in PCL
          retlw   'A'           ;return the ASCII
                                ; char A
          retlw   'B'           ;return the ASCII
                                ; char B
          retlw   'C'           ;return the ASCII
                                ; char C
          .
          .
          .
```

As I have already stated, these will be highly appreciated as you work with the PICMICRO® microcontroller line. Read tables have a variety of uses, and the deeper into this subject you get, the more often they will come to mind. In essence, they are one space-saving way to include information in your programs that can be, or needs to be, accessed repeatedly. I included this chapter because I really believe you will like read/lookup tables.

CHAPTER 7
THE PICMETER™ APPLICATION: A RESISTANCE/ CAPACITANCE MEASURING SYSTEM

INTRODUCTION

This is undoubtedly the most complex of the projects in this book, but also one of the most instructive and productive. Microchip application note AN611 (*Figures 7.1* and *7.2*), by Rodger Richey of Logic Products Division, was my guide for this one, and if you enjoy hobby electronics, I think you will enjoy this device. It provides a simple but accurate way to measure the value of both resistors and capacitors, and displays the results on the screen of a personal computer.

I say this is the most complex project, but it really isn't all that bad. The hardware is just a little more involved than some of the other devices, and the software is somewhat longer, but none of it is anything that should give you a problem. So, don't be intimidated by the PICMETER™ application. Dive in and have some fun!

This system utilizes another 18-pin PICMICRO® microcontroller, the PIC16C622, which has some extra

This device provides a simple but accurate way to measure the value of both resistors and capacitors, and displays the results on the screen of a personal computer.

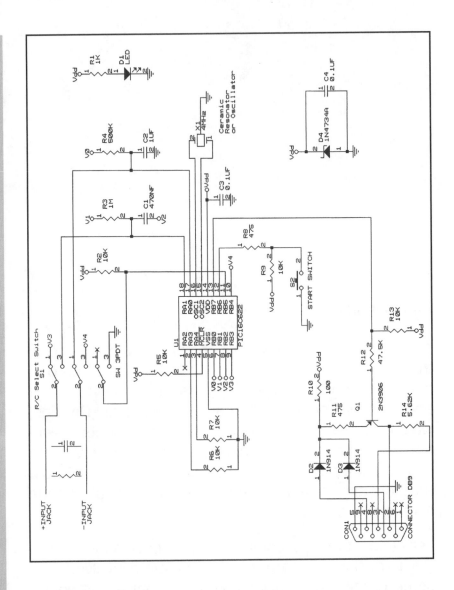

Figure 7.1. Schematic diagram for the PIC16C622-based resistance/capacitance meter.

goodies built in. For example, this chip offers two on-board analog comparators and an on-chip voltage reference. The value of both these features will become more than apparent as you navigate through this project. Additionally, the PIC16C622 has a program memory of

2K x 14 and a data memory of 128 x 8. This extra storage space will come in mighty handy.

Naturally, the PIC16C622 retains the admirable features of the other PIC16CXXX family, such as 13 I/O pins, PORTB interrupt, and 8-bit timer/counter with an 8-bit prescaler. Again, some of these characteristics will be employed in the PICMETER™.

As a sidebar, the PIC16C62X family also has a "brown-out" detector and a master clear (MCLR) filter. Both are useful in providing circuit stability. The brown-out will detect a positive voltage drop below a certain level (4 volts) and holds the chip in "reset" during that low-voltage situation. The filter is valuable in smoothing out glitches on the MCLR line.

Okay, let's take a good look at how this meter works. The easiest way to describe the PICMETER™ as we go along is to discuss the hardware and software together. I know this is a departure from some of the other chapters, but in this case it works. So, put on the proverbial thinking cap, because here we go!

THEORY OF THE PICMETER™ APPLICATION

During this discussion, from time to time, I will refer to Figure 7.1. This is, of course, the schematic diagram of the hardware. If you take a quick look at Figure 7.1, you will see a relatively simple circuit. There are a few more resistors, capacitors and/or other discrete components than are normally associated with PICMICRO® MCU projects, but the design is clean and straightforward. It

As a sidebar, the PIC16C62X family also has a "brown-out" detector and a master clear (MCLR) filter.

The PICMETER™ utilizes a modification of a tried and true method—the "single slope"-integrating converter.

will also easily fit on a single, rather small, printed circuit board (PCB).

At the center of the meter is the PIC16C622 microcontroller in a comparatively normal configuration. That is, connection of the input/output (I/O) ports and a 4-megahertz ceramic resonator as the clock source. Two switches are used for control, with S1 selecting the type of component to be tested (capacitor or resistor) and S2 acting as the "GO" button to initiate the test. The light-emitting diode (D1) is an "ON" indicator, while Zener diode (D4) regulates both the voltage on the PIC16C622 and the potential on the serial connection RTS and DTR lines to 5.6 volts.

Capacitors C3 and C4 are used to filter both the microcontroller and the Zener diode. Considering the PIC16C622 and company use a mere 7 milliamps of current, powering it from the serial port itself is not a problem. In fact, this is a relatively common technique in computer peripherals.

To further explain the switch operation, let's start with S1. Using the PORTB RB5 line, S1 selects the component type and also connects the unknown part to the resistor/capacitor (R/C) "tank" circuits (R3/C1 and R4/C2). The significance of this will later become clear. Switch S2, which is connected to PORTB RB6, initiates a measurement by using the PORTB "wake-up on change" interrupt. This tells the PIC16C622 a measurement is requested. If you are interested in a more in-depth look at the S2 procedure, it is a variation of Microchip application note AN552 that is available from the Microchip web site.

The PICMETER™ utilizes a modification of a tried and true method—the "single slope"-integrating converter. In an effort to explain how this works, the device compares an unknown voltage against an internal linear value. At a certain point in the measurement process, a comparator will change state (go from high to low or vice versa), and the time it takes for that to happen is equivalent to the value of the unknown input. This, of course, is compared to the time interval of the linear value.

Hence, with the PICMETER™, the charge time of the linear or known component is compared with the charge time of an unknown component, and the difference in these two intervals indicates the value of the unknown component. This may seem a little confusing, but trust me, it works quite well.

Timer0 is used to measure the charge time of the unknown RC network, and this value is multiplied by the known value. The result is stored in an accumulator. Next, the known RC network is timed and the accumulator is divided by this value. The result is the values of the "test" capacitor or resistor. *Figure 7.3* illustrates the formulas used to determine both the resistance and capacitance values. This is the "unknown resistance/capacitance" (R/CUNK) equals "unknown time" (tUNK) multiplied by "R/C known" (R/CKN) divided by "known time" (tKN). A fairly simple equation:

R/CUKN = tUNK x R/CKN/tKN

The result of the measurement is sent to the PC to be displayed on its screen. The PICMETER™ is capable of making resistance evaluation between 1 ohm and 999,000

RESISTANCE EQUATION

$$R_{UNK} = \frac{\tau_{UNK} \times R_{KN}}{\tau_{KN}}$$

CAPACITANCE EQUATION

$$C_{UNK} = \frac{\tau_{UNK} \times C_{KN}}{\tau_{KN}}$$

MATH ROUTINE ACCUMULATORS

Name	Operation	Result	Temp. Storage
Add	ACCa + ACCb	ACCb	N/A
Subtract	2's Comp ACCa then	ACCa	N/A
	ACCa + ACCb	ACCa	
Multiply	ACCa x ACCb	ACCb (MSB's) ACCc (LSB's)	ACCd
Divide	ACCb:ACCc ACCa	quotient in ACCc remainder in ACCb	ACCd
2's Comp	NOT(ACCa) + 1	ACCa	N/A

Figure 7.3. Resistor/capacitor equations and accumulator math routines.

ohms, while capacitance tests cover the 1 nanofarad to 999,000-picofarad range. And, that illustrates the merit this device can have on your test bench.

If you try to measure a resistance without a resistor connected to the test points, an error message will appear on the PC screen. This indicates the PICMETER™ is reading infinite resistance and knows that is not possible.

Also, a 0 picofarad reading in the capacitance mode will occur if no capacitor is connected to the test points. This is indicative of there being absolutely no capacitance present, again an almost impossible condition.

For a more software-related view of the measuring process, let me run through the following description. To start out, the I/O pins have to be reconfigured. In default, PORTA and PORTB are grounded outputs. Since the RC networks are connected to these lines, the capacitors are discharged. The first measurement is of the "unknown" component, so the "known" part, either R4 or C1, is removed from the R/C network by changing either the PIC16C622's RB0 or RB2 ports to an input. The other network is kept grounded.

Now, you initialize the analog modules, and the mode is set to "Four Inputs to Two Multiplexed Comparators". Voltage IN to comparator 1 is then selected as RA0 by clearing the CIS bit (CMCON<3>), and voltage IN to comparator 2 becomes RA1. Next, three things happen:

1) The high range is established.
2) The voltage reference is enabled.
3) The output is disabled.

The resistor ladder is set to "tap 12," because this is the lowest value of the reference voltage (VREF) that will activate the comparator and still provide a long enough time constant to produce a suitable measurement resolution.

The comparator flag is then cleared following a 20-microsecond delay to allow the analog modules to stabilize, and the comparator interrupts are enabled. Next,

If you try to measure a resistance without a resistor connected to the test points, an error message will appear on the PC screen. This indicates the PICMETER™ is reading infinite resistance and knows that is not possible.

the Timer0 is cleared, and as a last action, the PEIE bit is set to enable those interrupts, and the GIE bit is set to enable other interrupts.

With the analog modules in order, Timer0 is again cleared and power is applied to the unknown network by setting either RB1 or RB3 high. This results in three registers, which are cascaded together, to be incremented, and the charge time of the unknown component will be stored in these registers. While this is going on (waiting for the DONE flag), the ERROR flag is checked. The specifics of this will be covered shortly.

Timer0 is blocked from further incrementing the time registers when the capacitor voltage causes a comparator "state change." At that point, the DONE flag is set, and the charge time value in the time registers equals "tUNK". The analog system is then disabled, as well as the comparator interrupts, and the comparators are turned off. Now, ports RA0 to RA3 and RB0 to RB4 discharge the R/C network capacitors by again being set as grounded outputs.

The purpose here is to avert false results during the next reading. Additionally, to conserve power, the voltage reference is disabled. The interrupt flags are cleared and extra time delays are added to allow the capacitors to completely discharge.

A known value of either capacitance or resistance is now multiplied by the unknown time (tUNK). For reliability, the ohm or picofarad value of the known components is measured with a reliable test meter, and each value will be represented by a 24-bit number. The multiplied value will be 56 bits and will be stored in accumulator ACCb for the most significant 24 bits (MSB), and accumulator ACCc for the least significant 24 bits (LSB).

By making the PIC16C622 ports inputs, the unknown part is removed from the R/C network, and the charge time of the known R/C network is measured. Again, the analog systems are initialized and the same procedure as above is repeated. With the known R/C network value obtained, the 56-bit unknown product stored in accumulators ACCb and ACCc are divided by the known result. This process will yield another 24-bit value that will be designated in ohms or picofarads, depending on the unknown component type. This is then transmitted to the PC via a cable.

So, by now, you should have a fairly solid perception of how the measuring process is accomplished. As I said, it isn't a difficult procedure, but it can be a little confusing until you think it through. Anyway, I hope I didn't get too redundant. Next, let's talk about how the transmission to the personal computer is accomplished.

RS-232 TRANSMISSION

As might be expected, the PICMETER™ transfers information to the PC in a serial fashion. Using "transmit only," the meter employs software adopted from the Microchip application note AN593. Again, if you want or need additional information regarding this routine, check the Microchip web site.

Considering that the serial port is a product of the software, it is necessary to disable all interrupts during information

The measuring process isn't a difficult procedure, but it can be a little confusing until you think it through.

transfer. This prevents corruption of the baud rate. As soon as power is applied to the PICMETER™, a message is sent to the PC. It will read "PICMETER™ Booted!". A 4-bit packet structure is used, with 1 bit as the command bit followed by 3 bits of data. As for the command bit, it has to contain one of four commands as follows.

1) ASCII'S' signifies that a measurement has been initiated.
2) ASCII'E' tells the PC that an error has been detected.
3) ASCII'R' tells the PC that resistance data is contained in the three data bytes.
4) ASCII'C' tells the PC that capacitance data is contained in the three data bytes.

With either the resistance or capacitance commands, the first data byte contains the most significant bits (MSB) of the measured value, while the last data byte contains the least significant bits (LSB) of the measured value. Also, at this point in time, the three data bytes for both "S" and "E" do not contain useful information.

Each time you start the PICMETER™ by pressing switch S2, an "S" command is sent to the PC. Following that, unless there is an error "E", an "R" or "C" is sent for a valid measurement.

One of the nice features of the PICMETER™ is that it operates from a single supply.

One of the nice features of the PICMETER™ is that it operates from a single supply, but since that is the case, a discrete transistor (Q1) is employed as a "level shifter." This guarantees that a low value of between −3 volts and −11 volts appears on the RS-232 TXD line. When this line, the PIC16C622's RB7, is logic high, Q1 turns off, bringing the PC's RXD to approximately the same level as the TXD, or −3 to −11 volts.

A logic low on the PIC16C622 RB7 line turns Q1 on, however, and that brings the RXD line to near the same level as the DTR or RTS lines (approximately +3 to +11 volts). So, there are five pins of interest with a DB9 connector (CON1), and these are:

PIN 2 — RXD
PIN 3 — TXD
PIN 4 — DTR
PIN 5 — GND
PIN 7 — RTS

The RTS, DTR, and GND lines power the PICMETER™, while RXD is connected to Q1's collector. The TXD line is hooked to the RXD line through resistor R14, and since there is not any hardware "handshaking" involved, pin 6 (DSR) and pin 8 (CST) are left floating (disconnected).

As promised, let me take a closer look at the error message. At the risk of sounding simplistic, the error message is issued whenever the PIC16C622 detects an error. I know that is stating the obvious, but this becomes important. For example, the range for resistance is 1 ohm to 999,000 ohms, and if you are checking a resistor that doesn't have markings (I have bought surplus resistors that were perfectly good, but they were not banded) or the markings have faded from heat or have been otherwise deemed unreadable, the PICMETER™ will aid you in determining said resistor's value.

Since you do not have an idea of the resistor value, however, you will want to be advised if you exceed the range of the PICMETER™. Using C2 as the charging capacitor, resistor measurements should fall in a charge time between one millisecond and 999 milliseconds. If that time surmounts the 999 milliseconds, the PC screen is going to indicate an error. The same will happen if the charging time is less than one millisecond. Hence, the error message is telling you the resistor under test is outside the limits of the meter.

This is also true for capacitor measurements. And here, the likelihood of testing unmarked components is even greater. (Over the years, I have procured many an unmarked surplus capacitor). With R3 as part of the R/C network, the range is between 1 nanofarad and 999,000 picofarads, or a charging time of one millisecond to 999 milliseconds. Again, if these parameters are exceeded in one direction or the other, the PICMETER™ will send the computer an error message.

For a little more detail, the 4-megahertz resonator provides Timer0 with a resolution of one microsecond. This limits the time registers to a maximum count of 999,000, which is a 20-bit number. If the PC should receive a 21st bit, then something is wrong (either the test component is out of range or there is no component at all at the test points). A third possibility would be that S1 is incorrectly set for component "type" (for instance, trying to measure a capacitor in the resistance setting). And, you guessed it, if this happens, out comes the error message.

A quick word about the "math routines." These are depicted in Figure 7.3 and are the result of simple algorithms found in most texts on computer math. These include subtraction, division, multiplication, and 2's compliment, and are located in the PIC16C622's general random access memory (RAM). That area contains four accumulator sections—ACCa, ACCb, ACCc, and ACCd—for this purpose.

PERSONAL COMPUTER PROGRAM

Next comes the program used by your personal computer (PC) to display the PICMETER™ results. This is Microsoft Visual Basic code designed for a Windows platform. *Figure 7.4* shows what you will (should) see on your monitor, and the software for this program is listed in Figure 7.2, Appendix B.

With that out of the way, let's look at what this code does. The program is quite simple in nature and, quoting from the application notes, operates as follows.

1) Select the appropriate COM port by clicking on either the COM1 or COM2 buttons.

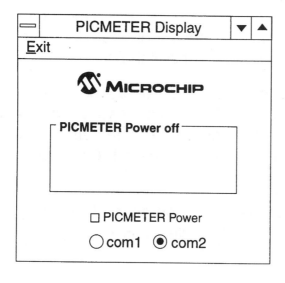

*Figure 7.4.
Personal computer display for
PICMETER™.*

2) Turn power on to the PICMETER™ by clicking on the PICMETER™ Power button.

3) The frame message should read "PICMETER Booted!", the frame contents will be cleared, and the LED on the PICMETER™ should be on.

4) The switch S1 selects the type of component that is in the measuring terminals.

5) Pressing the START button, S2, on the PICMETER™ will initiate a measurement. The frame message should read "Measuring Component," and the contents of the frame will be cleared.

6) When the measurement is complete, the frame message will read "Resistance" or "Capacitance" depending on the position of switch S1. The value of the component will be displayed in the frame as well as the units.

7) If an error is detected, the frame message will read "Error Detected". This is only a measurement error. Check the component on the measuring terminals and the position of switch S1.

8) Turn off the PICMETER™ by clicking on the PICMETER™ Power button. The frame message will change to "PICMETER Power OFF", the frame contents will be cleared, and the LED on the PICMETER™ will turn off.

So, as can be seen, operation of the PICMETER™ is very simple. And, while I talked a lot about checking unknown resistors and capacitors, don't forget this device can be used to check the value of *known* components as well. Just because a resistor says it is 100 ohms doesn't necessarily mean it is 100 ohms. Tolerance can be 5 percent, 10 percent, or even 20 percent, depending on the color of the fourth band (gold = 5%, silver = 10%, no band = 20%), and it is often prudent to confirm the actual component value before you solder it to a printed circuit board (PCB).

Okay! *Figure 7.5* shows two charts that represent the accuracy factor of the PICMETER™. You will notice that

The PICMETER measures capacitance in the range of 1 nF to 999 nF. Table 3 shows a comparison of various capacitors. All capacitors have a tolerance of 10% and have various dielectrics. The average error percentage is 3%.

The resistance range of the PICMETER is 1 kΩ to 999 kΩ. Table 4, Resistance Measurements, shows a comparison of various resistors in this range. All resistors have a tolerance of 5%. The average error percentage is 1%.

Figure 7.5.
Accuracy charts
for PICMETER™.

CAPACITANCE MEASUREMENTS

Capacitance Accuracy			
Marked Value	Fluke Value	PICMETER Value	Error %
2.2 nF	2.3 nF	2.2 nF	4.3
2.5 nF	2.63 nF	2.5 nF	4.9
20 nF	16.5 nF	16.3 nF	1.2
33 nF	35.2 nF	35.8 nF	1.7
47 nF	45 nF	44.5 nF	1.1
50 nF	52 nF	52.9 nF	1.7
100 nF	99.7 nF	93 nF	6.7
0.1 µF	95 nF	96.1 nF	1.2
0.1 µF	99.4 nF	102.8 nF	3.4
0.22 µF	215 nF	215.2 nF	0.1
470 nF	508 nF	518.9 nF	2.1
940 nF	922 nF	983.1 nF	6.6

The 2.5 nF, 100 nF and 940 nF capacitors all have polyester dielectric material. The Equivalent Series Resistance (ESR) of polyester capacitors is typically high which would cause the PICMETER to have a larger error than other dielectrics. If the error percentages for these capacitors is ignored, the average error decreases to 1.9%.

RESISTANCE MEASUREMENTS

Resistance Accuracy			
Marked Value	Fluke Value	PICMETER Value	Error %
1.2K	1.215K	1.2K	1.3
5.1K	5.05K	5.0K	1.0
8.2K	8.47K	8.3K	2.0
10K	10.2K	10K	2.0
15K	15.36K	15.1K	1.7
20K	20.8K	20.5K	1.5
30K	30.4K	30K	1.4
51K	50.3K	49.8K	1.0
75K	75.5K	74.8K	1.0
91K	96.4K	95.9K	0.6
150K	146.3K	145.6K	0.5
200K	195.5K	195K	0.3
300K	309K	309.5K	0.2
430K	433K	434.5K	0.4
560K	596K	599.6K	0.6
680K	705K	709.8K	0.7
820K	901K	907.3K	0.7
910K	970K	977.8K	0.8

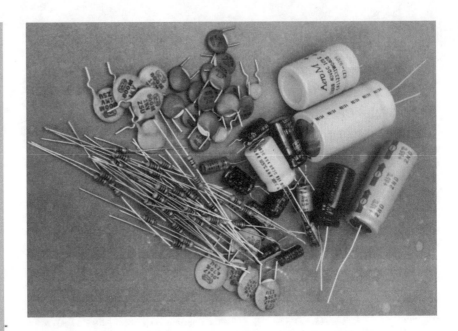

Figure 7.6.
A variety of
resistors and
capacitors—the
components that
can be checked
with the com-
pleted project in
this chapter.

the PICMETER™ well exceeds the standard tolerances for both capacitors and resistors. Hence, you can take the results you get to the bank. Well, you could do that, but I don't know what you will gain by doing so.

Never mind! The important issue here is the PICMETER™ very accurately measures both resistors and capacitors (see *Figure 7.6*) as long as they fall within its measurement range.

Referring to Figure 7.5, you will note that the average tolerance for capacitors is in the 10% range, while the average PICMETER™ percentage of error is 3%. Respectively, the average error percentage for resistance values is 1%, and that equals even the "precision" variety of commercial resistor. Thus, it's hard to complain about the reading you will get with this gem.

CONCLUSION

In conclusion … is that redundant? Well, maybe so, but you have to cut me some slack—you know, literary license. Anyway, in conclusion, I think you will like this device. Not only is it a terrific training experience, it adds a very pragmatic test instrument to you workbench arsenal. I don't know about you, but I need all the help I can get in that area.

Construction is very simple, and while the code is a little long, it is easy to enter into the PIC16C622, and the product of your labor is well worth the effort. I know what you're saying, "He says that about everything!" But, once you discover the convenience of designing around PICMICRO® microcontrollers, especially in this case, you are going to agree with me. Mark my words! Have fun, and enjoy!

Not only is this project a terrific training experience, it adds a very pragmatic test instrument to you workbench arsenal.

AN611

PLEASE CHECK MICROCHIP TECHNOLOGY'S WEBSITE FOR THE LATEST VERSION OF THE SOURCE CODE.

APPENDIX A: PICMETER FIRMWARE

```
MPASM 01.02.05 Intermediate PICMETER.ASM   5-1-1995 11:29:17                PAGE 1
PICMETER Firmware for PIC16C622

LOC OBJECT CODE    LINE SOURCE TEXT
VALUE

                   0001        TITLE "PICMETER Firmware for PIC16C622"
                   0002        LIST P = 16C622, F = INHX8M
                   0003
                   0004        INCLUDE "C:\PICMASTR\P16CXX.INC"
                   0002 ; P16CXX.INC  Standard Header File, Version 0.2 Microchip Technology, Inc.
                   0004
                   0005
3FB9               0006        FUSES _BODEN_OFF&_CP_OFF&_PWDT_ON&_WDT_OFF&_XT_OSC
                   0007
                   0008 ;*********************************************************************
                   0009 ;*-------------------------------------------------------------------*
                   0010 ;*-                                                                 -*
                   0011 ;*-    PICMETER - Resistance and Capacitance Meter                  -*
                   0012 ;*-                                                                 -*
                   0013 ;*-------------------------------------------------------------------*
                   0014 ;*-                                                                 -*
                   0015 ;*-    Author:      Rodger Richey                                   -*
                   0016 ;*-                 Applications Engineer                           -*
                   0017 ;*-    Filename:    picmtr.asm                                      -*
                   0018 ;*-    Revision:    1 May 1995                                      -*
                   0019 ;*-                                                                 -*
                   0020 ;*-------------------------------------------------------------------*
                   0021 ;*-                                                                 -*
                   0022 ;*-    PICMETER is based on a PIC16C622 which has two comparators and -*
                   0023 ;*-    a variable voltage reference. Resistance and capacitance is  -*
                   0024 ;*-    calculated by measuring the time constant of a RC network. The -*
                   0025 ;*-    toggle switch selects either resistor or capacitor input. The -*
                   0026 ;*-    pushbutton switch starts a measurement. The time constant of the -*
                   0027 ;*-    unknown component is compared to that of known component to  -*
                   0028 ;*-    calculate the value of the unknown component.  The following -*
                   0029 ;*-    formulas are used:                                          -*
                   0030 ;*-                                                                 -*
                   0031 ;*-    Resistance:    Ru = ( Rk * Tu ) / Tk                        -*
                   0032 ;*-    Capacitance:   Cu = ( Ck * Tu ) / Tk                        -*
                   0033 ;*-                                                                 -*
                   0034 ;*-------------------------------------------------------------------*
                   0035 ;*********************************************************************
                   0036
                   0037
                   0038 ;*********************************************************************
                   0039 ;*-------------------------------------------------------------------*
                   0040 ;*-    RS232 code borrowed from Application Note AN593              -*
                   0041 ;*-    "Serial Port Routines Without Using the RTCC"               -*
                   0042 ;*-    Author: Stan D'Souza                                        -*
                   0043 ;*-------------------------------------------------------------------*
                   0044 ;*********************************************************************
003D 0900          0045 xtal        equ    .4000000
2580               0046 baud        equ    .9600
000F 4240          0047 fclk        equ    xtal/4
                   0048 ;*********************************************************************
                   0049 ;The value baudconst must be a 8-bit value only
0020               0050 baudconst        equ    ((fclk/baud)/3-2)
                   0051 ;*********************************************************************
                   0052
                   0053
```

Figure 7.2. *(Continued on next 14 pages)* **Code for the PIC16C622-based resistance/capacitance meter.**

AN611

```
                    0054 ;*********************************************************************
                    0055 ;         Bit Equates
                    0056 ;*********************************************************************
0000                0057 BEGIN    equ    0              ;begin a measurement flag
0007                0058 DONE     equ    7              ;done measuring flag
0005                0059 WHICH    equ    5              ;R or C measurement flag
0003                0060 F_ERROR  equ    3              ;error detection flag
0005                0061 EMPTY    equ    5              ;flag if component is connected
0000                0062 V0       equ    0              ;power for R reference ckt
0001                0063 V1       equ    1              ;power for C reference ckt
0002                0064 V2       equ    2              ;ground for C reference ckt
0003                0065 V3       equ    3              ;power for unknown R ckt
0004                0066 V4       equ    4              ;ground for unknown C ckt
0007                0067 msb_bit  equ    7              ;define for bit 7
0000                0068 lsb_bit  equ    0              ;define for bit 0
0007                0069 RkHI     equ    0x07           ;value of the known resistance, R4, in ohms
009D                0070 RkMID    equ    0x9D           ;measured by a Fluke meter
0038                0071 RkLO     equ    0x38
0007                0072 CkHI     equ    0x07           ;value of the known capacitance, C1, in pF
00C8                0073 CkMID    equ    0xC8           ;measured by a Fluke meter
0030                0074 CkLO     equ    0x30
                    0075
                    0076 ;*********************************************************************
                    0077 ;         User Registers
                    0078 ;*********************************************************************
                    0079 ;         Bank 0
0020                0080 W_TEMP      equ    0x20         ;Bank 0 temporary storage for W reg
0021                0081 STATUS_TEMP equ 0x21           ;temporary storage for STATUS reg
0023                0082 Ttemp    equ    0x23           ;temporary Time register
0024                0083 flags    equ    0x24           ;flags register
0025                0084 count    equ    0x25           ;RS232 register
0026                0085 txreg    equ    0x26           ;RS232 data register
0027                0086 delay    equ    0x27           ;RS232 delay register
0028                0087 offset   equ    0x28           ;table position register
0029                0088 msb      equ    0x29           ;general delay register
002A                0089 lsb      equ    0x2A           ;general delay register
0040                0090 TimeLO   equ    0x40           ;Time registers
0041                0091 TimeMID  equ    0x41
0042                0092 TimeHI   equ    0x42
                    0093
                    0094 ;         Math related registers
0050                0095 ACCaHI   equ    0x50           ;24-Bit accumulator a
0051                0096 ACCaMID  equ    0x51
0052                0097 ACCaLO   equ    0x52
0053                0098 ACCbHI   equ    0x53           ;24-Bit accumulator b
0054                0099 ACCbMID  equ    0x54
0055                0100 ACCbLO   equ    0x55
0056                0101 ACCcHI   equ    0x56           ;24-Bit accumulator c
0057                0102 ACCcMID  equ    0x57
0058                0103 ACCcLO   equ    0x58
0059                0104 ACCdHI   equ    0x59           ;24-Bit accumulator d
005A                0105 ACCdMID  equ    0x5A
005B                0106 ACCdLO   equ    0x5B
005C                0107 temp     equ    0x5C           ;temporary storage
                    0108
                    0109 ;         User Registers Bank 1
                    0110 ;W_TEMP   equ    0xA0           ;Bank 1 temporary storage for W reg
                    0111
                    0112 ;         User defines
                    0113 #define tx        PORTB,7       ;define for RS232 TXD output pin
                    0114
                    0115 ;*********************************************************************
                    0116
                    0117          org    0x0
0000 2810           0118          goto   init
                    0119
```

AN611

```
              0120          org     0x4
0004 28B9     0121          goto    ServiceInterrupts
              0122
              0123          org     0x10
0010          0124 init
0010 1283     0125          bcf     STATUS,RP0      ;select bank 0
0011 0185     0126          clrf    PORTA           ;clear PORTA and PORTB
0012 0186     0127          clrf    PORTB
0013 1786     0128          bsf     tx              ;set TXD output pin
0014 01A4     0129          clrf    flags           ;clear flags register
0015 3010     0130          movlw   0x10            ;load table offset register
0016 00A8     0131          movwf   offset
0017 018B     0132          clrf    INTCON          ;clear interrupt flags and disable interrupts
0018 3007     0133          movlw   0x07            ;turn off comparators, mode 111
0019 009F     0134          movwf   CMCON
001A 2140     0135          call    delay20         ;wait for comarators to settle
001B 089F     0136          movf    CMCON,F
001C 130C     0137          bcf     PIR1,CMIF
001D 1683     0138          bsf     STATUS,RP0      ;select bank 1
001E 3088     0139          movlw   0x88            ;WDT prescalar,internal TMR0 increment
001F 0081     0140          movwf   OPTION_REG
0020 0185     0141          clrf    TRISA           ;PORTA all outputs, discharges RC ckts
0021 3060     0142          movlw   0x60            ;PORTA<7,4:0> outputs, PORTA<6:5> inputs
0022 0086     0143          movwf   TRISB
0023 300C     0144          movlw   0x0C            ;setup Voltage Reference
0024 009F     0145          movwf   VRCON
0025 1283     0146          bcf     STATUS,RP0      ;select bank 0
0026 3008     0147          movlw   0x08            ;enable RBIE interrupt
0027 008B     0148          movwf   INTCON
0028 213D     0149          call    vlong           ;delay before transmitting boot message
0029 213D     0150          call    vlong           ;to allow computer program to setup
002A 213D     0151          call    vlong
002B 2131     0152          call    BootMSG         ;transmit boot message
002C 178B     0153          bsf     INTCON,GIE      ;enable global interrupt bit
              0154
002D          0155 start
002D 1C24     0156          btfss   flags,BEGIN     ;wait for a start measurement key press
002E 282D     0157          goto    start
002F 1024     0158          bcf     flags,BEGIN     ;clear start measurement flag
              0159
0030 138B     0160          bcf     INTCON,GIE      ;transmit a start measurement message
0031 3053     0161          movlw   'S'             ;to the PC
0032 20AD     0162          call    Send
0033 178B     0163          bsf     INTCON,GIE
              0164
0034 01C2     0165          clrf    TimeHI          ;reset Time registers
0035 01C1     0166          clrf    TimeMID
0036 01C0     0167          clrf    TimeLO
0037 1E86     0168          btfss   PORTB,WHICH     ;detect if resistor or capacitor measure
0038 2862     0169          goto    Capacitor
              0170
0039          0171 Resistor
0039 1683     0172          bsf     STATUS,RP0      ;set V0 to input
003A 1406     0173          bsf     TRISB,V0
003B 1283     0174          bcf     STATUS,RP0
003C 20FB     0175          call    AnalogOn        ;turn analog on
003D 0181     0176          clrf    TMR0
003E 0000     0177          nop
003F 1586     0178          bsf     PORTB,V3        ;turn power on to unknown RC ckt
0040 19A4     0179 RwaitU   btfsc   flags,F_ERROR   ;detect if an error occurs
0041 288B     0180          goto    ErrorDetect
0042 1FA4     0181          btfss   flags,DONE      ;measurement completed flag
0043 2840     0182          goto    RwaitU
0044 13A4     0183          bcf     flags,DONE      ;clear measurement completed flag
0045 2111     0184          call    AnalogOff       ;turn analog off
              0185
```

AN611

```
0046 2126    0186          call    SwapTtoA      ;swap Time to accumulator a
0047 3007    0187          movlw   RkHI          ;swap known resistance value
0048 00D3    0188          movwf   ACCbHI        ;to accumulator b
0049 309D    0189          movlw   RkMID
004A 00D4    0190          movwf   ACCbMID
004B 3038    0191          movlw   RkLO
004C 00D5    0192          movwf   ACCbLO
004D 2230    0193          call    Mpy24         ;multiply accumulator a and b
             0194
004E 1683    0195          bsf     STATUS,RP0    ;set V3 to input
004F 1586    0196          bsf     TRISB,V3
0050 1283    0197          bcf     STATUS,RP0
0051 20FB    0198          call    AnalogOn      ;turn analog on
0052 0181    0199          clrf    TMR0
0053 0000    0200          nop
0054 1406    0201          bsf     PORTB,V0      ;turn power on to known RC ckt
0055 19A4    0202 RwaitK   btfsc   flags,F_ERROR ;detect if an error occurs
0056 288B    0203          goto    ErrorDetect
0057 1FA4    0204          btfss   flags,DONE    ;measurement completed flag

0058 2855    0205          goto    RwaitK
0059 13A4    0206          bcf     flags,DONE    ;clear measurement completed flag
005A 2111    0207          call    AnalogOff     ;turn analog off
             0208
005B 2126    0209          call    SwapTtoA      ;swap Time to accumulator a
005C 224B    0210          call    Div24         ;divide multiply by known time
             0211
005D 138B    0212          bcf     INTCON,GIE    ;disable all interrupts
005E 3052    0213          movlw   'R'           ;transmit, for R measurement
005F 20AD    0214          call    Send
0060 178B    0215          bsf     INTCON,GIE    ;enable global interrupt bit
0061 282D    0216          goto    start         ;restart
             0217
0062         0218 Capacitor
0062 1683    0219          bsf     STATUS,RP0    ;set V2 to input
0063 1506    0220          bsf     TRISB,V2
0064 1283    0221          bcf     STATUS,RP0
0065 20FB    0222          call    AnalogOn      ;turn analog on
0066 0181    0223          clrf    TMR0
0067 0000    0224          nop
0068 1486    0225          bsf     PORTB,V1      ;turn power on to unknown RC ckt
0069 19A4    0226 CwaitU   btfsc   flags,F_ERROR ;detect if an error occurs
006A 288B    0227          goto    ErrorDetect
006B 1FA4    0228          btfss   flags,DONE    ;measurement completed flag
006C 2869    0229          goto    CwaitU
006D 13A4    0230          bcf     flags,DONE    ;clear measurement completed flag
006E 2111    0231          call    AnalogOff     ;turn analog off
             0232
006F 2126    0233          call    SwapTtoA      ;swap Time to accumulator a
0070 3007    0234          movlw   CkHI          ;swap known resistance value
0071 00D3    0235          movwf   ACCbHI        ;to accumulator b
0072 30C8    0236          movlw   CkMID
0073 00D4    0237          movwf   ACCbMID
0074 3030    0238          movlw   CkLO
0075 00D5    0239          movwf   ACCbLO
0076 2230    0240          call    Mpy24         ;multiply accumulator a and b
             0241
0077 1683    0242          bsf     STATUS,RP0    ;set V3 to input
0078 1606    0243          bsf     TRISB,V4
0079 1283    0244          bcf     STATUS,RP0
007A 20FB    0245          call    AnalogOn      ;turn analog on
007B 0181    0246          clrf    TMR0
007C 0000    0247          nop
007D 1486    0248          bsf     PORTB,V1      ;turn power on to known RC ckt
007E 19A4    0249 CwaitK   btfsc   flags,F_ERROR ;detect if an error occurs
007F 288B    0250          goto    ErrorDetect
```

AN611

```
0080 1FA4     0251        btfss    flags,DONE      ;measurement completed flag
0081 287E     0252        goto     CwaitK
0082 13A4     0253        bcf      flags,DONE      ;clear measurement completed flag
0083 2111     0254        call     AnalogOff       ;turn analog off
              0255
0084 2126     0256        call     SwapTtoA        ;swap Time to accumulator a
0085 224B     0257        call     Div24           ;divide multiply by known time
              0258
0086 138B     0259        bcf      INTCON,GIE      ;disable all interrupts
0087 3043     0260        movlw    'C'             ;transmit, for C measurement
0088 20AD     0261        call     Send
0089 178B     0262        bsf      INTCON,GIE      ;enable global interrupt bit
008A 282D     0263        goto     start           ;restart
              0264
008B          0265 ErrorDetect
008B 1283     0266        bcf      STATUS,RP0      ;disable TMR0
008C 128B     0267        bcf      INTCON,T0IE
008D 110B     0268        bcf      INTCON,T0IF
008E 2111     0269        call     AnalogOff       ;turn analog off
008F 11A4     0270        bcf      flags,F_ERROR   ;clear error flag
              0271
0090 138B     0272        bcf      INTCON,GIE      ;disable all interrupts
0091 3045     0273        movlw    'E'             ;transmit, for C measurement
0092 20AD     0274        call     Send
0093 178B     0275        bsf      INTCON,GIE      ;enable global interrupt bit
0094 282D     0276        goto     start           ;restart
              0277
              0278 ;****************************************************************
              0279 ;*--------------------------------------------------------------*
              0280 ;*-    RS232 Transmit Routine
              0281 ;*-    Borrowed from AN593, "Serial Port Routines Without Using the RTCC"
              0282 ;*-    Author: Stan D'Souza
              0283 ;*-    This is the routine that interfaces directly to the hardware
              0284 ;*--------------------------------------------------------------*
              0285 ;****************************************************************
0095          0286 Transmit
0095 1283     0287        bcf      STATUS,RP0
0096 00A6     0288        movwf    txreg
0097 1386     0289        bcf      tx              ;send start bit
0098 3020     0290        movlw    baudconst
0099 00A7     0291        movwf    delay
009A 3009     0292        movlw    0x9
009B 00A5     0293        movwf    count
009C          0294 txbaudwait
009C 0BA7     0295        decfsz   delay
009D 289C     0296        goto     txbaudwait
009E 3020     0297        movlw    baudconst
009F 00A7     0298        movwf    delay
00A0 0BA5     0299        decfsz   count
00A1 28A6     0300        goto     SendNextBit
00A2 3009     0301        movlw    0x9
00A3 00A5     0302        movwf    count
00A4 1786     0303        bsf      tx              ;send stop bit
00A5 0008     0304        return
00A6          0305 SendNextBit
00A6 0CA6     0306        rrf      txreg
00A7 1C03     0307        btfss    STATUS,C
00A8 28AB     0308        goto     Setlo
00A9 1786     0309        bsf      tx
00AA 289C     0310        goto     txbaudwait
00AB 1386     0311 Setlo  bcf      tx
00AC 289C     0312        goto     txbaudwait
              0313 ;_____
              0314
              0315 ;****************************************************************
              0316 ;*--------------------------------------------------------------*
```

AN611

```
                  0317 ;*-      Generic Transmit Routine
                  0318 ;*-      Sends what is currently in the W register and accumulator ACCc
                  0319 ;*------------------------------------------------------------------*
                  0320 ;******************************************************************
00AD              0321 Send
00AD 2095         0322         call    Transmit
00AE 2146         0323         call    delay1                  ;delay between bytes
00AF 0856         0324         movf    ACCcHI,W                ;transmit high resistance byte
00B0 2095         0325         call    Transmit
00B1 2146         0326         call    delay1                  ;delay between bytes
00B2 0857         0327         movf    ACCcMID,W               ;transmit mid resistance byte
00B3 2095         0328         call    Transmit
00B4 2146         0329         call    delay1                  ;delay between bytes
00B5 0858         0330         movf    ACCcLO,W                ;transmit low resistance byte
00B6 2095         0331         call    Transmit
00B7 2146         0332         call    delay1                  ;delay between bytes
00B8 0008         0333         return
                  0334 ;_____
                  0335
                  0336 ;******************************************************************
                  0337 ;*------------------------------------------------------------------*
                  0338 ;*-      Interrupt Service Routines
                  0339 ;*------------------------------------------------------------------*
                  0340 ;******************************************************************
00B9              0341 ServiceInterrupts
00B9 00A0         0342         movwf   W_TEMP                  ;Pseudo push instructions
00BA 0E03         0343         swapf   STATUS,W
00BB 1283         0344         bcf     STATUS,RP0
00BC 00A1         0345         movwf   STATUS_TEMP
                  0346
00BD 0801         0347         movf    TMR0,W
00BE 00A3         0348         movwf   Ttemp
00BF 190B         0349         btfsc   INTCON,T0IF             ;Service Timer 0 overflow
00C0 20E5         0350         call    ServiceTimer
00C1 1B0C         0351         btfsc   PIR1,CMIF               ;Stops Timer0, Records Value
00C2 20EC         0352         call    ServiceComparator
00C3 180B         0353         btfsc   INTCON,RBIF             ;Service pushbutton switch
00C4 20CB         0354         call    ServiceKeystroke        ;Starts a measurement
                  0355
00C5 1283         0356         bcf     STATUS,RP0
00C6 0E21         0357         swapf   STATUS_TEMP,W           ;Pseudo pop instructions
00C7 0083         0358         movwf   STATUS
00C8 0EA0         0359         swapf   W_TEMP,F
00C9 0E20         0360         swapf   W_TEMP,W
                  0361
00CA 0009         0362         retfie
                  0363 ;_____
                  0364
                  0365 ;******************************************************************
                  0366 ;*------------------------------------------------------------------*
                  0367 ;*-      Borrowed from AN552, "Implementing Wake-up on Key Stroke"
                  0368 ;*-      Author: Stan D'Souza
                  0369 ;*------------------------------------------------------------------*
                  0370 ;******************************************************************
00CB              0371 ServiceKeystroke
00CB 118B         0372         bcf     INTCON,RBIE             ;disable interrupt
00CC 0906         0373         comf    PORTB,W                 ;read PORTB
00CD 100B         0374         bcf     INTCON,RBIF             ;clear interrupt flag
00CE 3940         0375         andlw   B'01000000'
00CF 1903         0376         btfsc   STATUS,Z
00D0 28D6         0377         goto    NotSwitch
00D1 2143         0378         call    delay16                 ;de-bounce switch for 16msec
00D2 0906         0379         comf    PORTB,W                 ;read PORTB again
00D3 20D9         0380         call    KeyRelease              ;check for key release
00D4 1424         0381         bsf     flags,BEGIN
00D5 0008         0382         return
```

AN611

```
                        0383
00D6                    0384 NotSwitch                         ;detected other PORTB pin change
00D6 100B               0385        bcf     INTCON,RBIF        ;reset RBI interrupt
00D7 158B               0386        bsf     INTCON,RBIE
00D8 0008               0387        return
                        0388
00D9                    0389 KeyRelease
00D9 2143               0390        call    delay16            ;debounce switch
00DA 0906               0391        comf    PORTB,W            ;read PORTB
00DB 100B               0392        bcf     INTCON,RBIF        ;clear flag
00DC 158B               0393        bsf     INTCON,RBIE        ;enable interrupt
00DD 3940               0394        andlw   B'01000000'
00DE 1903               0395        btfsc   STATUS,Z           ;key still pressed?
00DF 0008               0396        return                     ;if no, then return
00E0 0063               0397        sleep                      ;else, save power
00E1 118B               0398        bcf     INTCON,RBIE        ;disable interrupts
00E2 0906               0399        comf    PORTB,W            ;read PORTB
00E3 100B               0400        bcf     INTCON,RBIF        ;clear flag
00E4 28D9               0401        goto    KeyRelease         ;try again
                        0402 ;_____
                        0403
                        0404 ;************************************************************************
                        0405 ;*--------------------------------------------------------------------*
                        0406 ;*-     ISR to service a Timer0 overflow
                        0407 ;*--------------------------------------------------------------------*
                        0408 ;************************************************************************
00E5                    0409 ServiceTimer
00E5 0AC1               0410        incf    TimeMID,F          ;increment middle Time byte
00E6 1903               0411        btfsc   STATUS,Z           ;if middle overflows,
00E7 0AC2               0412        incf    TimeHI,F           ;increment high Time byte
00E8 1AC2               0413        btfsc   TimeHI,EMPTY       ;check if component is connected
00E9 15A4               0414        bsf     flags,F_ERROR      ;set error flag
00EA 110B               0415        bcf     INTCON,T0IF        ;clear TMR0 interrupt flag
00EB 0008               0416        return
                        0417 ;_____
                        0418
                        0419 ;************************************************************************
                        0420 ;*--------------------------------------------------------------------*
                        0421 ;*-     ISR to service a Comparator interrupt
                        0422 ;*--------------------------------------------------------------------*
                        0423 ;************************************************************************
00EC                    0424 ServiceComparator
00EC 1283               0425        bcf     STATUS,RP0         ;select bank 0
00ED 1E86               0426        btfss   PORTB,WHICH        ;detect which measurement, R or C?
00EE 28F2               0427        goto    capcomp
00EF 1F1F               0428        btfss   CMCON,C1OUT        ;detect if R ckt has interrupted
00F0 28F4               0429        goto    scstop
00F1 28F9               0430        goto    scend
00F2                    0431 capcomp
00F2 1B9F               0432        btfsc   CMCON,C2OUT        ;detect if C ckt has interrupted
00F3 28F9               0433        goto    scend
00F4                    0434 scstop
00F4 128B               0435        bcf     INTCON,T0IE        ;disable TMR0 interrupts
00F5 110B               0436        bcf     INTCON,T0IF
00F6 0823               0437        movf    Ttemp,W
00F7 00C0               0438        movwf   TimeLO
00F8 17A4               0439        bsf     flags,DONE         ;set DONE flag
00F9                    0440 scend
00F9 130C               0441        bcf     PIR1,CMIF          ;clear comparator interrupt flag
00FA 0008               0442        return
                        0443 ;_____
                        0444
                        0445 ;************************************************************************
                        0446 ;*--------------------------------------------------------------------*
                        0447 ;*-     Turn Comparators and Vref On
                        0448 ;*--------------------------------------------------------------------*
```

AN611

```
                  0449 ;**************************************************************
00FB              0450 AnalogOn
00FB 1283         0451        bcf     STATUS,RP0       ;select bank 0
00FC 3002         0452        movlw   0x02             ;turn comparators on, mode 010
00FD 009F         0453        movwf   CMCON            ;4 inputs multiplexed to 2 comparators
00FE 1683         0454        bsf     STATUS,RP0       ;select bank 1
00FF 300F         0455        movlw   0x0F             ;make PORTA<3:0> all inputs
0100 0085         0456        movwf   TRISA
0101 179F         0457        bsf     VRCON,VREN
0102 1283         0458        bcf     STATUS,RP0       ;select bank 0
0103 2140         0459        call    delay20          ;20msec delay
0104 089F         0460        movf    CMCON,F          ;clear comparator mismatch condition
0105 130C         0461        bcf     PIR1,CMIF        ;clear comparator interrupt flag
0106 1683         0462        bsf     STATUS,RP0
0107 170C         0463        bsf     PIE1,CMIE        ;enable comparator interrupts
0108 1283         0464        bcf     STATUS,RP0
0109 170B         0465        bsf     INTCON,PEIE      ;enable peripheral interrupts
010A 11A4         0466        bcf     flags,F_ERROR
010B 0181         0467        clrf    TMR0             ;clear TMR0 counter
010C 0000         0468        nop
010D 0000         0469        nop
010E 110B         0470        bcf     INTCON,T0IF      ;clear TMR0 interrupt flag
010F 168B         0471        bsf     INTCON,T0IE      ;enable TMR0 interrupts
0110 0008         0472        return
                  0473 ;_____
                  0474
                  0475 ;**************************************************************
                  0476 ;*----------------------------------------------------------*
                  0477 ;*-     Turn Comparators and Vref Off
                  0478 ;*----------------------------------------------------------*
                  0479 ;**************************************************************
0111              0480 AnalogOff
0111 1283         0481        bcf     STATUS,RP0
0112 130B         0482        bcf     INTCON,PEIE
0113 3080         0483        movlw   0x80             ;reset PORTB value
0114 0086         0484        movwf   PORTB
0115 1683         0485        bsf     STATUS,RP0       ;select bank 1
0116 130C         0486        bcf     PIE1,CMIE        ;disable comparator interrupts
0117 0185         0487        clrf    TRISA            ;set PORTA pins to outputs, discharge RC ckt
0118 3060         0488        movlw   0x60             ;set PORTB 7,4-0 as outputs, 6,5 as inputs
0119 0086         0489        movwf   TRISB
011A 139F         0490        bcf     VRCON,VREN       ;disable Vref
011B 1283         0491        bcf     STATUS,RP0       ;select bank 0
011C 3007         0492        movlw   0x07
011D 009F         0493        movwf   CMCON            ;disable comparators
011E 2140         0494        call    delay20          ;20msec delay
011F 089F         0495        movf    CMCON,F          ;clear comparator mismatch condition
0120 130C         0496        bcf     PIR1,CMIF        ;clear comparator interrupt flag
0121 110B         0497        bcf     INTCON,T0IF      ;clear Timer0 interrupt flag
0122 213D         0498        call    vlong            ;long delay to allow capacitors to discharge
0123 213D         0499        call    vlong
0124 213D         0500        call    vlong
0125 0008         0501        return
                  0502 ;_____
                  0503
                  0504 ;**************************************************************
                  0505 ;*----------------------------------------------------------*
                  0506 ;*-     Swap Time to Accumulator a
                  0507 ;*----------------------------------------------------------*
                  0508 ;**************************************************************
0126              0509 SwapTtoA
0126 1283         0510        bcf     STATUS,RP0
0127 0842         0511        movf    TimeHI,W
0128 00D0         0512        movwf   ACCaHI
0129 0841         0513        movf    TimeMID,W
012A 00D1         0514        movwf   ACCaMID
```

AN611

```
012B 0840        0515          movf      TimeLO,W
012C 00D2        0516          movwf     ACCaLO
012D 01C2        0517          clrf      TimeHI
012E 01C1        0518          clrf      TimeMID
012F 01C0        0519          clrf      TimeLO
0130 0008        0520          return
                 0521 ;_____
                 0522
                 0523 ;*****************************************************************
                 0524 ;*---------------------------------------------------------------*
                 0525 ;*-     Transmit the Boot Message
                 0526 ;*---------------------------------------------------------------*
                 0527 ;*****************************************************************
0131             0528 BootMSG
0131 1283        0529          bcf       STATUS,RP0      ;select bank 0
0132 3002        0530 msg      movlw     HIGH Table      ;init the PCH for a table call
0133 008A        0531          movwf     PCLATH
0134 0828        0532          movf      offset,W        ;move table offset into W
0135 2200        0533          call      Table           ;get table value
0136 2095        0534          call      Transmit        ;transmit table value
0137 2146        0535          call      delay1          ;delay between bytes
0138 0BA8        0536          decfsz    offset,F        ;check for end of table
0139 2932        0537          goto      msg
013A 3010        0538          movlw     0x10            ;reset table offset
013B 00A8        0539          movwf     offset
013C 0008        0540          return
                 0541 ;_____
                 0542
                 0543 ;*****************************************************************
                 0544 ;*---------------------------------------------------------------*
                 0545 ;*-     Delay Routines
                 0546 ;*---------------------------------------------------------------*
                 0547 ;*****************************************************************
013D 30FF        0548 vlong    movlw     0xff            ;very long delay, approx 200msec
013E 00A9        0549          movwf     msb
013F 2948        0550          goto      d1
0140             0551 delay20                            ;20 msec delay
0140 301A        0552          movlw     .26
0141 00A9        0553          movwf     msb
0142 2948        0554          goto      d1
0143             0555 delay16                            ;16 msec delay
0143 3015        0556          movlw     .21
0144 00A9        0557          movwf     msb
0145 2948        0558          goto      d1
0146             0559 delay1                             ;approx 750nsec delay
0146 3001        0560          movlw     .1
0147 00A9        0561          movwf     msb
0148 30FF        0562 d1       movlw     0xff
0149 00AA        0563          movwf     lsb
014A 0BAA        0564 d2       decfsz    lsb,F
014B 294A        0565          goto      d2
014C 0BA9        0566          decfsz    msb,F
014D 2948        0567          goto      d1
014E 0008        0568          return
                 0569 ;_____
                 0570
                 0571
                 0572          org       0x200
                 0573
                 0574
                 0575 ;*****************************************************************
                 0576 ;*---------------------------------------------------------------*
                 0577 ;*-     Table for Boot Message
                 0578 ;*---------------------------------------------------------------*
                 0579 ;*****************************************************************
0200             0580 Table                              ;boot message "PICMETER Booted!"
```

AN611

```
0200 0782    0581        addwf    PCL                ;add W to PCL
0201 3400    0582        retlw    0
0202 3421    0583        retlw    '!'
0203 3464    0584        retlw    'd'
0204 3465    0585        retlw    'e'
0205 3474    0586        retlw    't'
0206 346F    0587        retlw    'o'
0207 346F    0588        retlw    'o'
0208 3442    0589        retlw    'B'
0209 3420    0590        retlw    ' '
020A 3472    0591        retlw    'r'
020B 3465    0592        retlw    'e'
020C 3474    0593        retlw    't'
020D 3465    0594        retlw    'e'
020E 346D    0595        retlw    'm'
020F 3443    0596        retlw    'C'
0210 3449    0597        retlw    'I'
0211 3450    0598        retlw    'P'
             0599 ;_____
             0600
             0601 ;******************************************************************
             0602 ;*----------------------------------------------------------------*
             0603 ;*-    24-bit Addition
             0604 ;*-
             0605 ;*-    Uses ACCa and ACCb
             0606 ;*-
             0607 ;*-    ACCa + ACCb -> ACCb
             0608 ;*----------------------------------------------------------------*
             0609 ;******************************************************************
0212         0610 Add24
0212 0852    0611        movf     ACCaLO,W
0213 07D5    0612        addwf    ACCbLO             ;add low bytes
0214 1803    0613        btfsc    STATUS,C           ;add in carry if necessary
0215 2A1D    0614        goto     A2
0216 0851    0615 A1     movf     ACCaMID,W
0217 07D4    0616        addwf    ACCbMID            ;add mid bytes
0218 1803    0617        btfsc    STATUS,C           ;add in carry if necessary
0219 0AD3    0618        incf     ACCbHI
021A 0850    0619        movf     ACCaHI,W
021B 07D3    0620        addwf    ACCbHI             ;add high bytes
021C 3400    0621        retlw    0
021D 0AD4    0622 A2     incf     ACCbMID
021E 1903    0623        btfsc    STATUS,Z
021F 0AD3    0624        incf     ACCbHI
0220 2A16    0625        goto     A1
             0626 ;_____
             0627
             0628 ;******************************************************************
             0629 ;*----------------------------------------------------------------*
             0630 ;*-    Subtraction ( 24 - 24 -> 24 )
             0631 ;*-
             0632 ;*-    Uses ACCa, ACCb, ACCd
             0633 ;*-
             0634 ;*-    ACCa -> ACCd,
             0635 ;*-    2's complement ACCa,
             0636 ;*-    call Add24 ( ACCa + ACCb -> ACCb ),
             0637 ;*-    ACCd -> ACCa
             0638 ;*----------------------------------------------------------------*
             0639 ;******************************************************************
0221         0640 Sub24
0221 0850    0641        movf     ACCaHI,W           ;Transfer ACCa to ACCd
0222 00D9    0642        movwf    ACCdHI
0223 0851    0643        movf     ACCaMID,W
0224 00DA    0644        movwf    ACCdMID
0225 0852    0645        movf     ACCaLO,W
0226 00DB    0646        movwf    ACCdLO
```

AN611

```
0227 2275    0647         call    compA          ;2's complement ACCa
0228 2212    0648         call    Add24          ;Add ACCa to ACCb
0229 0859    0649         movf    ACCdHI,W       ;Transfer ACCd to ACCa
022A 00D0    0650         movwf   ACCaHI
022B 085A    0651         movf    ACCdMID,W
022C 00D1    0652         movwf   ACCaMID
022D 085B    0653         movf    ACCdLO,W
022E 00D2    0654         movwf   ACCaLO
022F 3400    0655         retlw   0
             0656 ;_____
             0657
             0658 ;****************************************************************
             0659 ;*--------------------------------------------------------------*
             0660 ;*-     Multiply ( 24 X 24 -> 56 )
             0661 ;*-
             0662 ;*-     Uses ACCa, ACCb, ACCc, ACCd
             0663 ;*-
             0664 ;*-     ACCa * ACCb -> ACCb,ACCc   56-bit output
             0665 ;*-     with ACCb (ACCbHI,ACCbMID,ACCbLO) with 24 msb's and
             0666 ;*-     ACCc (ACCcHI,ACCcMID,ACCcLO) with 24 lsb's
             0667 ;*--------------------------------------------------------------*
             0668 ;****************************************************************
0230         0669 Mpy24
0230 223F    0670         call    Msetup
0231 0CD9    0671 mloop   rrf     ACCdHI         ;rotate d right
0232 0CDA    0672         rrf     ACCdMID
0233 0CDB    0673         rrf     ACCdLO
0234 1803    0674         btfsc   STATUS,C       ;need to add?
0235 2212    0675         call    Add24
0236 0CD3    0676         rrf     ACCbHI
0237 0CD4    0677         rrf     ACCbMID
0238 0CD5    0678         rrf     ACCbLO
0239 0CD6    0679         rrf     ACCcHI
023A 0CD7    0680         rrf     ACCcMID
023B 0CD8    0681         rrf     ACCcLO
023C 0BDC    0682         decfsz  temp           ;loop until all bits checked
023D 2A31    0683         goto    mloop
023E 3400    0684         retlw   0
             0685
023F         0686 Msetup
023F 3018    0687         movlw   0x18           ;for 24 bit shifts
0240 00DC    0688         movwf   temp
0241 0853    0689         movf    ACCbHI,W       ;move ACCb to ACCd
0242 00D9    0690         movwf   ACCdHI
0243 0854    0691         movf    ACCbMID,W
0244 00DA    0692         movwf   ACCdMID
0245 0855    0693         movf    ACCbLO,W
0246 00DB    0694         movwf   ACCdLO
0247 01D3    0695         clrf    ACCbHI
0248 01D4    0696         clrf    ACCbMID
0249 01D5    0697         clrf    ACCbLO
024A 3400    0698         retlw   0
             0699 ;_____
             0700
             0701 ;****************************************************************
             0702 ;*--------------------------------------------------------------*
             0703 ;*-     Division ( 56 / 24 -> 24 )
             0704 ;*-
             0705 ;*-     Uses ACCa, ACCb, ACCc, ACCd
             0706 ;*-
             0707 ;*-     56-bit dividend in ACCb,ACCc ( ACCb has msb's and ACCc has lsb's)
             0708 ;*-     24-bit divisor in ACCa
             0709 ;*-     quotient is stored in ACCc
             0710 ;*-     remainder is stored in ACCb
             0711 ;*--------------------------------------------------------------*
             0712 ;****************************************************************
```

AN611

```
024B              0713 Div24
024B 2272         0714        call   Dsetup
                  0715
024C 1003         0716 dloop  bcf    STATUS,C
024D 0DD8         0717        rlf    ACCcLO          ;Rotate dividend left 1 bit position
024E 0DD7         0718        rlf    ACCcMID
024F 0DD6         0719        rlf    ACCcHI
0250 0DD5         0720        rlf    ACCbLO
0251 0DD4         0721        rlf    ACCbMID
0252 0DD3         0722        rlf    ACCbHI
                  0723
0253 1803         0724        btfsc  STATUS,C        ;invert carry and exclusive or with the
0254 2A58         0725        goto   clear           ;msb of the divisor then move this bit
0255 1FD0         0726        btfss  ACCaHI,msb_bit  ;into the lsb of the dividend
0256 0AD8         0727        incf   ACCcLO
0257 2A5A         0728        goto   cont
0258 1BD0         0729 clear  btfsc  ACCaHI,msb_bit
0259 0AD8         0730        incf   ACCcLO
                  0731
025A 1858         0732 cont   btfsc  ACCcLO,lsb_bit  ;check the lsb of the dividend
025B 2A5E         0733        goto   minus
025C 2212         0734        call   Add24           ;if = 0, then add divisor to upper 24 bits
025D 2A5F         0735        goto   check           ;of dividend
025E 2221         0736 minus  call   Sub24           ;if = 1, then subtract divisor from upper
                  0737                                ;24 bits of dividend
                  0738
025F 0BDC         0739 check  decfsz temp,f          ;do 24 times
0260 2A4C         0740        goto   dloop
                  0741
0261 1003         0742        bcf    STATUS,C
0262 0DD8         0743        rlf    ACCcLO          ;shift lower 24 bits of dividend 1 bit
0263 0DD7         0744        rlf    ACCcMID         ;position left
0264 0DD6         0745        rlf    ACCcHI
0265 1BD3         0746        btfsc  ACCbHI,msb_bit  ;exlusive or the inverse of the msb of the
0266 2A6A         0747        goto   w1              ;dividend with the msb of the divisor
0267 1FD0         0748        btfss  ACCaHI,msb_bit  ;store in the lsb of the dividend
0268 0AD8         0749        incf   ACCcLO
0269 2A6C         0750        goto   wzd
026A 1BD0         0751 w1     btfsc  ACCaHI,msb_bit
026B 0AD8         0752        incf   ACCcLO
026C 1FD3         0753 wzd    btfss  ACCbHI,msb_bit  ;if the msb of the remainder is set and
026D 2A71         0754        goto   wend
026E 1BD0         0755        btfsc  ACCaHI,msb_bit  ;the msb of the divisor is not
026F 2A71         0756        goto   wend
0270 2212         0757        call   Add24           ;add the divisor to the remainder to correct
                  0758                                ;for zero partial remainder
                  0759
0271 3400         0760 wend   retlw  0               ;quotient in 24 lsb's of dividend
                  0761                                ;remainder in 24 msb's of dividend
                  0762
0272              0763 Dsetup
0272 3018         0764        movlw  0x18            ;loop 24 times
0273 00DC         0765        movwf  temp
                  0766
0274 3400         0767        retlw  0
                  0768 ;_____
                  0769
                  0770 ;****************************************************************
                  0771 ;*--------------------------------------------------------------*
                  0772 ;*-      2's Complement
                  0773 ;*-
                  0774 ;*-      Uses ACCa
                  0775 ;*-
                  0776 ;*-      Performs 2's complement conversion on ACCa
                  0777 ;*--------------------------------------------------------------*
                  0778 ;****************************************************************
```

AN611

```
0275                 0779 compA
0275 09D2            0780        comf   ACCaLO       ;invert all bits in accumulator a
0276 09D1            0781        comf   ACCaMID
0277 09D0            0782        comf   ACCaHI
0278 0AD2            0783        incf   ACCaLO       ;add one to accumulator a
0279 1903            0784        btfsc  STATUS,Z
027A 0AD1            0785        incf   ACCaMID
027B 1903            0786        btfsc  STATUS,Z
027C 0AD0            0787        incf   ACCaHI
027D 3400            0788        retlw  0
                     0789 ;_____
                     0790
                     0791        END
                     0792

0000 : X---X---------- XXXXXXXXXXXXXXXX XXXXXXXXXXXXXXXX XXXXXXXXXXXXXXXX
0040 : XXXXXXXXXXXXXXXX XXXXXXXXXXXXXXXX XXXXXXXXXXXXXXXX XXXXXXXXXXXXXXXX

0080 : XXXXXXXXXXXXXXXX XXXXXXXXXXXXXXXX XXXXXXXXXXXXXXXX XXXXXXXXXXXXXXXX
00C0 : XXXXXXXXXXXXXXXX XXXXXXXXXXXXXXXX XXXXXXXXXXXXXXXX XXXXXXXXXXXXXXXX

0100 : XXXXXXXXXXXXXXXX XXXXXXXXXXXXXXXX XXXXXXXXXXXXXXXX XXXXXXXXXXXXXXXX
0140 : XXXXXXXXXXXXXXX- ---------------- ---------------- ----------------

0200 : XXXXXXXXXXXXXXXX XXXXXXXXXXXXXXXX XXXXXXXXXXXXXXXX XXXXXXXXXXXXXXXX
0240 : XXXXXXXXXXXXXXXX XXXXXXXXXXXXXXXX XXXXXXXXXXXXXXXX XXXXXXXXXXXXX--

All other memory blocks unused.


Errors    :    0
Warnings  :    0
Messages  :    0
```

AN611

PLEASE CHECK MICROCHIP TECHNOLOGY'S WEBSITE FOR THE LATEST VERSION OF THE SOURCE CODE.

APPENDIX B: VISUAL BASIC PROGRAM

PICMTR.FRM

```
Sub Form_Load ()
      'Initialize the program
      Image1.Height = 600
      Image1.Width = 2700
      Frame1.Caption = "PICMETER Power Off"
      Label1.Caption = ""
      Label2.Caption = ""

      'Initialize Comm Port 1
      Comm1.RThreshold = 1
      Comm1.Handshaking = 0
      Comm1.Settings = "9600,n,8,1"
      Comm1.CommPort = 2
      Comm1.PortOpen = True

      'Initialize the global variable First%
      First% = 0
End Sub

Sub Form_Unload (Cancel As Integer)
      'Unload PICMETER
      Comm1.RTSEnable = False
      Comm1.DTREnable = False
      Comm1.PortOpen = False
      Unload PICMETER
End Sub

Sub Comm1_OnComm ()
      Dim Value As Double
      Dim High As Double
      Dim Medium As Double
      Dim Low As Double

      'Received a character
      If Comm1.CommEvent = 2 Then
          If First% = 0 Then
              If Comm1.InBufferCount = 16 Then
                   Label1.FontSize = 10
                   InString$ = Comm1.Input
                   If InString$ = "PICMETER Booted!" Then
                        Frame1.Caption = "PICMETER Booted!"
                   End If
                   First% = 1
                   Comm1.InputLen = 4
              End If
          Else
              If Comm1.InBufferCount >= 4 Then
                   InString$ = Comm1.Input
                   If Left$(InString$, 1) = "R" Then
                        Frame1.Caption = "Resistance"
                        Label2.FontName = "Symbol"
                        Label2.Caption = "KW"
                        Label1.FontSize = 24
                   ElseIf Left$(InString$, 1) = "C" Then
                        Frame1.Caption = "Capacitance"
                        Label2.FontName = "MS Sans Serif"
                        Label2.Caption = "nF"
                        Label1.FontSize = 24
                   ElseIf Left$(InString$, 1) = "E" Then
                        Frame1.Caption = "Error Detected"
                        Label2.Caption = ""
                   ElseIf Left$(InString$, 1) = "S" Then
                        Frame1.Caption = "Measuring Component"
```

AN611

```
                    Label2.Caption = ""
                Else
                    Frame1.Caption = "Error Detected"
                    Label2.Caption = ""
                End If

                If Frame1.Caption = "Error Detected" Then
                    Label1.Caption = ""
                ElseIf Frame1.Caption = "Measuring Component" Then
                    Label1.Caption = ""
                Else
                    High = 65536# * Asc(Mid$(InString$, 2, 1))
                    Medium = 256# * Asc(Mid$(InString$, 3, 1))
                    Low = Asc(Mid$(InString$, 4, 1))
                    Label1.Caption = Format$((High + Medium + Low) / 1000, "###0.0")
                End If
            End If
        End If
    End If
End Sub

Sub Check3D1_Click (Value As Integer)
    'Control Power to the PICMETER
    If Check3D1.Value = False Then
        Comm1.InputLen = 0
        Label1.Caption = ""
        Label2.Caption = ""
        Comm1.RTSEnable = False
        Comm1.DTREnable = False
        Frame1.Caption = "PICMETER Power Off"
        InString$ = Comm1.Input
    Else
        Frame1.Caption = ""
        First% = 0
        Comm1.InputLen = 0
        InString$ = Comm1.Input
        Comm1.RTSEnable = True
        Comm1.DTREnable = True
    End If
End Sub

Sub menExitTop_Click ()
    'Unload PICMETER
    Unload PICMETER
End Sub

Sub Option1_Click ()
    'Open COM1 for communications
    If Option1.Value = True Then
        If Comm1.CommPort = 2 Then
            Comm1.PortOpen = False
            Comm1.CommPort = 1
            Comm1.PortOpen = True
        End If
    End If
End Sub

Sub Option2_Click ()
    'Open COM2 for communications
    If Option2.Value = True Then
        If Comm1.CommPort = 1 Then
            Comm1.PortOpen = False
            Comm1.CommPort = 2
            Comm1.PortOpen = True
        End If
    End If
End Sub
```

PICMETER.BAS

```
Global I%
Global First%
```

CHAPTER 8

INTERFACING LCD MODULES TO PICMICRO® MCU PROJECTS

INTRODUCTION

One of the best aspects of using microcontrollers is their ability to directly drive the numerous liquid crystal display (LCD) modules available today. These displays come from many different companies, many different sources, and in many different sizes and configurations (see *Figure 8.4*). They are abundant on the surplus market, as well as with retail suppliers, and the price tag has really come into line in the last several years.

Thus, they make a terrific addition to many PICMICRO® microcontroller projects. Unlike their light-emitting diode (LED) cousins, however, they *do* require a controller, and the vast majority of these modules utilize the very efficient Hitachi HD44780A chip for that purpose. This makes things simple, as the 44780 will handle up to 80 characters and is highly versatile in its application. More will be said about this chip soon.

Liquid crystal display (LCD) modules make a terrific addition to many PICMICRO® microcontroller projects.

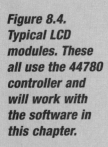

Figure 8.4. Typical LCD modules. These all use the 44780 controller and will work with the software in this chapter.

Interfacing the modules to a PICMICRO® MCU is quite easy (see *Figure 8.1*). As always, software will play a pivotal role in all this, and that makes the hardware just about as simple as it gets. A few power lines, control lines, and data lines are all that is required to have an LCD up and running. The Microchip application note AN587 (*Figures 8.1* and *8.3*) by Mark Palmer and Scott Fink was very helpful for this chapter, especially concerning the software.

Considering their ease of use and accessibility these days, LCD modules are a natural when it comes to displaying digital information. Furthermore, the performance of LCD modules is nothing short of impressive. Using very little

8-BIT DATA INTERFACE

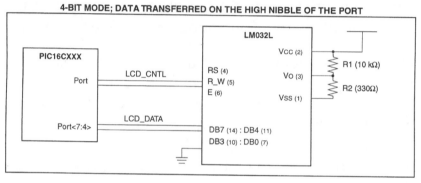

4-BIT MODE; DATA TRANSFERRED ON THE HIGH NIBBLE OF THE PORT

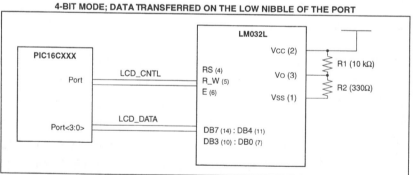

4-BIT MODE; DATA TRANSFERRED ON THE LOW NIBBLE OF THE PORT

Figure 8.1. Schematics for interfacing LCD modules to PICMICRO® MCUs. The LCD_CNTL and LCD_DATA lines are user definable to their port assignment. This is accomplished with EQUate statements in the source code.

power, the modules will conveniently display all kinds of stuff. So, next time you're planning a PICMICRO® MCU project, keep these gems in mind. Naturally, if a display is not required, this chapter won't help you. But, if one is needed, here's your boy.

THE HITACHI HD44780 LCD CONTROLLER

As promised, here is a little background on this marvelous chip. Virtually all of the LCD modules I have worked with have employed this controller or, in some rare occasions, an equivalent chip. I have heard about other controllers being used on the modules, but I have yet to encounter such chips. And, as I said, this makes everything very simple, as you are always dealing with the same operational configuration and power requirements.

Many of the modules have a single in-line pin (SIP) header or solder pad arrangement for connection to the LCD board. Each position associated with this layout is going to correspond to the same point on another module. That means that position 1 on one board and position 1 on another LCD will be the ground (Vss) connection. But, before I get myself into deep trouble, let me qualify that statement just a tad.

When I say "position 1," I'm referring to the first hookup point with respect to the HD44780 controller. In actuality, position 1 might be something else on other boards.

When I say "position 1," I'm referring to the first hookup point with respect to the HD44780 controller. In actuality, position 1 might be something else on other boards. For example, with the Ocular DM1621 display used in the prototype lab (see Chapter 2), position 1 is assigned to one side of the backlight (if one is present). Position 2 is assigned to the other side of the backlight. Hence, the actual controller connections don't start until position 3. Position 3 is the ground (Vss), which equates to position 1 on most other displays. Does that make sense? Sure it does! The best rule of thumb here is to acquire the data sheet for the display module you are using.

There is another variation on this style of connector. While the single in-line pin configuration seems to be the most popular, I have also seen modules with a dual in-line pin (DIP) arrangement. Here, two rows of seven connectors each are laid out side by side—that is, usually! In the case of the DM1621, there would be two rows of eight connectors. One row handles positions 1, 3, 5, 7, 9, 11, and 13, and the other row addresses 2, 4, 6, 8, 10, 12, and 14. Again, these can be a "header" or "solder pads," but either way, hookup to the modules becomes a very straightforward matter.

With that said, let's take a quick look at what each of the 14 connections is used for. As I said, with the exception of some strange and/or very unusual modules, these assignments are universal with the HD44780, and are as follows.

Position   Assignment

- Position 1   Ground (Vss)
- Position 2   Positive 5 Volts (Vdd)
- Position 3   Contrast Voltage (Vee)
- Position 4   Register Select (RS)
- Position 5   Read/Write (R/W)
- Position 6   Enable (E)
- Position 7   Data Bit DB0
- Position 8   Data Bit DB1
- Position 9   Data Bit DB2
- Position 10   Data Bit DB3
- Position 11   Data Bit DB4
- Position 12   Data Bit DB5
- Position 13   Data Bit DB6
- Position 14   Data Bit DB7

The first three locations are the power connections. These modules like to see a positive 5 volts and not much else. There is little room for variance from that specification, as too low a voltage will result in erratic behavior (sort of like Wiley Coyote) and, even worse, too high a voltage will burn the 44780 to a crisp. Not a good deal for your LCD module. If fact, it quickly becomes an expensive paperweight.

The third, or "contrast voltage" line (Vee) is sent to the "wiper" (the center connection) of a potentiometer. The ends of the potentiometer are wired across the ground (line 1) and the positive voltage (line 3), and that allows the wiper to vary the Vee potential to a level that provides the appropriate display contrast. This is usually in the negative 2.5-volt to positive 2.5-volt range.

The next three lines (4, 5, and 6) are used for "control." That is, they handle the "housekeeping" duties, or such things as whether to "read" or "write" data, and which "mode" to be in (data or command). The "enable," of course, enables the LCD display, and all of this is run by the software.

The last eight lines are the "data" lines. The 44780 can be operated in one of three different data modes, and the code determines which mode by setting two flags— the "Four-bit" flag and the "Data_HI" flag. These are set in the standard fashion of being either "ON" (1) or "OFF" (0), and result in the following operational modes.

FOUR_BIT	DATA_HI	MODE
1	0	4-bit mode, Data in low nibble
1	1	4-bit mode, Data in high nibble
0	X	8-bit mode

In an effort to explain what is going on here, many of the applications involving LCD modules with the HD44780 controller use only half the total number of data lines (7 to 14) to introduce data to the module. Hence, you will want to send the data to either the lower (LSB) section of the memory (lines 7 to 10) or to the higher (MSB) section of the memory (lines 11 to 14). More often than not the higher four bits, or "nibble," are used, and I don't really know why. By setting the above-mentioned flags, as illustrated in the chart, the software accomplishes this task. Naturally, if all eight data lines are used, you will want to employ the "8-bit mode."

Whether to use a 4-bit or 8-bit arrangement is, like so many things in electronics, a compromise (see *Figure 8.2*). The 4-bit modes require two data transfers per character or command, whereas the 8-bit mode only demands one. Hence, the 4-bit mode further slows down an already slow process. With the 8-bit mode, however, the hardware becomes more complex. You will now have to provide eight data lines from the PICMICRO® MCU to the LCD module. And, that can sometimes be problematic.

Another consideration involves the data-transfer speed. As stated, these displays tend to be fairly deliberate devices when compared to the microcontroller. So, it will be the job of the software to keep the LCD and PICMICRO® MCU in sync. Again, let me say that it is a very good idea to obtain a copy of the data sheet for the device you are using. But, often that will come with the module you order.

With many LCD units, the default mode is going to be 8 bits. If you are using a 4-bit mode, the last initialization

Whether to use a 4-bit or 8-bit arrangement is, like so many things in electronics, a compromise.

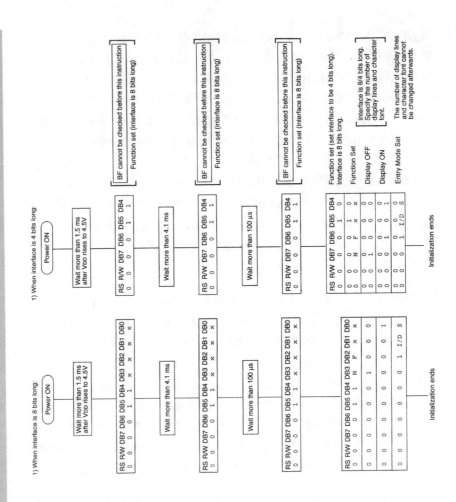

Figure 8.2.
Chart for initializ-
ing flow for LCD
modules.

command will have to specify the data-traffic rate (sometimes referred to as "width") to 4 bits. Additionally, a delay of about 4.6 milliseconds is required before the module can be initialized. This, of course, adds to the slow nature of the displays. Some typical LCD commands are as follows.

1) 1 or 2 lines of characters.
2) Display "on" or "off".

3) Load character address pointer.
4) Increment, or do not increment, character address pointer after each character.
5) Clear display.

As might be expected, these will appear in much of the code written for LCD modules.

Okay, that provides a quick overview of liquid crystal display modules. The code for this chapter will work well with any module using the HD44780A controller chip, but can easily be modified if an oddball controller should be encountered. All in all, the LCD units will make life much easier when it comes to adding information displays to your PICMICRO® MCU projects.

CONCLUSION

Well, there you have it: a simple, efficient, and *cheap* way to employ LCD modules in PICMICRO® microcontroller projects. Once you begin working with these beauties, you will learn to appreciate their practicality. Since there are only 14 interface lines—and that represents the most complex configuration—connecting the LCDs to a circuit is a snap.

If the project requires two separate boards (one for the electronics and one to hold the module), a header can be incorporated on the circuit board to accommodate the cable to the LCD. This often is a good idea, as it allows you to unplug the display board from the main board. That not only makes construction and installation more convenient, but it also protects against the individual connecting wires breaking.

And, if the project will be constructed on a single board, you will only have to run a maximum of 14 lines or traces to the LCD module. That modest number of connections can greatly simplify the layout, as opposed to running the higher number of lines often necessary with other systems.

So, as with all the neat stuff we are looking at in this book, keep the modules in mind for future endeavors into the microcontroller world. I'm a firm believer in making things as simple as possible when working with electronics (some people call it being lazy), and the LCD display modules certainly help achieve that goal. Additionally, they make working with the otherwise cumbersome liquid crystal display panels themselves a far easier experience. Yeah boy!

If the project requires two separate boards (one for the electronics and one to hold the module), a header can be incorporated on the circuit board to accommodate the cable to the LCD.

AN587

APPENDIX C: 4-BIT DATA INTERFACE, HIGH NIBBLE LISTING

```
MPASM 01.40.01 Intermediate   LM032L.ASM   4-7-1997  9:50:32        PAGE  1


LOC  OBJECT CODE     LINE  SOURCE TEXT
     VALUE

                     00001            LIST P=16C64
                     00002            ERRORLEVEL  -302
                     00003     ;
                     00004     ; This program interfaces to a Hitachi (LM032L) 2 line by 20 character display
                     00005     ; module. The program assembles for either 4-bit or 8-bit data interface, depending
                     00006     ; on the value of the 4bit flag. LCD_DATA is the port which supplies the data to
                     00007     ; the LM032L, while LCD_CNTL is the port that has the control lines ( E, RS, RW ).
                     00008     ; In 4-bit mode the data is transfer on the high nibble of the port ( PORT<7:4> ).
                     00009     ;
                     00010     ;         Program = LM032L.ASM
                     00011     ;         Revision Date:   5-10-94
                     00012     ;                          1-22-97    Compatibility with MPASMWIN 1.40
                     00013     ;
                     00014     ;
                     00015            include <p16c64.inc>
                     00001            LIST
                     00002     ; P16C64.INC  Standard Header File, Version 1.01    Microchip Technology, Inc.
                     00238            LIST
                     00016
0000009F             00017  ADCON1    EQU      9F
                     00018
00000000             00019  FALSE     EQU      0
00000001             00020  TRUE      EQU      1
                     00021
                     00022            include <lm032l.h>
                     00069            list
                     00023     ;
00000000             00024  Four_bit  EQU      FALSE           ; Selects 4- or 8-bit data transfers
00000001             00025  Data_HI   EQU      TRUE            ; If 4-bit transfers, Hi or Low nibble of PORT
                     00026     ;
                     00027            if ( Four_bit && !Data_HI )
                     00029     ;
00000000             00030  LCD_DATA       EQU      PORTB
00000001             00031  LCD_DATA_TRIS  EQU      TRISB
```

Figure 8.3.
(Continued on next 12 pages) **Code for LCD module interfacing to PICMICRO® MCUs.**

AN587

```
00032 ;   else
00033 ;
00034 ;
00000008          00035 LCD_DATA      EQU  PORTD
00000088          00036 LCD_DATA_TRIS EQU  TRISD
00037 ;   endif
00038 ;
00039 ;
00000005          00040 LCD_CNTL      EQU  PORTA
00041 ;
00042 ;
00043 ;
00044 ; LCD Display Commands and Control Signal names.
00045 ;
00046 ;     if ( Four_bit && !Data_HI )
00047 ;
00048 E    EQU  0          ; LCD Enable control line
00049 RW   EQU  1          ; LCD Read/Write control line
00050 RS   EQU  2          ; LCD Register Select control line
00051 ;
00052 ;   else
00053 ;
00000003          00054 E    EQU  3     ; LCD Enable control line
00000002          00055 RW   EQU  2     ; LCD Read/Write control line
00000001          00056 RS   EQU  1     ; LCD Register Select control line
00057 ;   endif
00058 ;
00059 ;
00060 ;
00000030          00061 TEMP1   EQU  0x030
00062 ;
0000 2808         00063         org    RESET_V        ; RESET vector location
                  00064 RESET   GOTO   START          ;
00065 ; This is the Periperal Interrupt routine. Should NOT get here
00066 ;
00067 ;
0004              00068         page   org
                  00069 PER_INT_V  ISR_V             ; Interrupt vector location
00070 ;
0004 1283         00071 ERROR1   BCF   STATUS, RP0    ; Bank 0
0006 1407         00072          BSF   PORTC, 0
0007 1007         00073          BCF   PORTC, 0
0007 2804         00074          GOTO  ERROR1
00075 ;
00076 ;
00077 ;
0008              00078 START                         ; POWER_ON Reset (Beginning of program)
```

AN587

```
0008 0183   00079           CLRF    STATUS              ; Do initialization (Bank 0)
0009 018B   00080           CLRF    INTCON
000A 018C   00081           CLRF    PIR1
000B 1683   00082           BSF     STATUS, RP0         ; Bank 1
000C 3000   00083           MOVLW   0x00                ; The LCD module does not like to work w/ weak pull-ups
000D 0081   00084           MOVWF   OPTION_REG
000E 018C   00085           CLRF    PIE1                ; Disable all peripheral interrupts
            00086   ;***
            00087   ;*** If using device with A/D, these two instructions are required.
            00088   ;***
            00089   ;
            00090           MOVLW   0xFF                ; Port A is Digital.
            00091           MOVWF   ADCON1
            00092   ;
000F 1283   00093           BCF     STATUS, RP0         ; Bank 0
0010 0185   00094           CLRF    PORTA               ; ALL PORT output should output Low.
0011 0186   00095           CLRF    PORTB
0012 0187   00096           CLRF    PORTC
0013 0188   00097           CLRF    PORTD
0014 0189   00098           CLRF    PORTE
0015 1010   00099           BCF     T1CON, TMR1ON       ; Timer 1 is NOT incrementing
            00100   ;
0016 1683   00101           BSF     STATUS, RP0         ; Select Bank 1
0017 0185   00102           CLRF    TRISA               ; RA5 - 0 outputs
0018 30F0   00103           MOVLW   0xF0                ; RB7 - 4 inputs, RB3 - 0 outputs
0019 0086   00104           MOVWF   TRISB
001A 0187   00105           CLRF    TRISC               ; RC Port are outputs
001B 1407   00106           BSF     TRISC, T1OSO        ; RC0 needs to be input for the oscillator to function
001C 0188   00107           CLRF    TRISD               ; RD Port are outputs
001D 0189   00108           CLRF    TRISE               ; RE Port are outputs
001E 140C   00109           BSF     PIE1, TMR1IE        ; Enable TMR1 Interrupt
001F 1781   00110           BSF     OPTION_REG,NOT_RBPU ; Disable PORTB pull-ups
0020 1283   00111           BCF     STATUS, RP0         ; Select Bank 0
            00112   ;
            00113           page
            00114   ;
            00115   ; Initilize the LCD Display Module
            00116   ;
0021 0185   00117           CLRF    LCD_CNTL            ; ALL PORT output should output Low.
            00118   ;
0022        00119   DISPLAY_INIT
            00120           if ( Four_bit && !Data_HI )
            00121           MOVLW   0x02                ; Command for 4-bit interface low nibble
            00122           endif
            00123   ;
            00124           if ( Four_bit && Data_HI )
            00125           MOVLW   0x020               ; Command for 4-bit interface high nibble
```

AN587

```
                  00126        endif
                  00127 ;
                  00128        if ( !Four_bit )
0022 3038         00129        MOVLW   0x038           ; Command for 8-bit interface
                  00130        endif
                  00131 ;
0023 0088         00132        MOVWF   LCD_DATA        ;
0024 1585         00133        BSF     LCD_CNTL, E     ;
0025 1185         00134        BCF     LCD_CNTL, E     ;
                  00135 ; This routine takes the calculated times that the delay loop needs to
                  00136 ; be executed, based on the LCD_INIT_DELAY EQUate that includes the
                  00137 ; frequency of operation. These use registers before they are needed to
                  00138 ; store the time.
                  00139 ;
                  00140 ;
0026 3006         00141 LCD_DELAY   MOVLW   LCD_INIT_DELAY  ; Use MSD and LSD Registers to Initilize LCD
0027 0B83         00142             MOVWF   MSD
0028 01B4         00143             CLRF    LSD
0029 0BB4         00144 LOOP2       DECFSZ  LSD, F          ; Delay time = MSD * ((3 * 256) + 3) * Tcy
002A 2829         00145             GOTO    LOOP2           ;
002B 0BB3         00146             DECFSZ  MSD, F          ;
002C 2829         00147 END_LCD_DELAY
                  00148             GOTO    LOOP2           ;
                  00149 ;
                  00150 ; Command sequence for 2 lines of 5x7 characters
                  00151 ;
002D              00152 CMD_SEQ
                  00153 ;
                  00154        if ( Four_bit )
                  00155        if ( !Data_HI )
                  00156        MOVLW   0X02            ; 4-bit low nibble xfer
                  00157        else
                  00158        MOVLW   0X020           ; 4-bit high nibble xfer
                  00159        endif
                  00160 ;
                  00161        else
002D 3038         00162        MOVLW   0X038           ; 8-bit mode
                  00163        endif
                  00164 ;
002E 0088         00165        MOVWF   LCD_DATA        ; This code for both 4-bit and 8-bit modes
002F 1585         00166        BSF     LCD_CNTL, E     ;
0030 1185         00167        BCF     LCD_CNTL, E     ;
                  00168 ;
                  00169        if ( Four_bit )         ; This code for only 4-bit mode (2nd xfer)
                  00170        if ( !Data_HI )
                  00171        MOVLW   0x08            ; 4-bit low nibble xfer
                  00172        else
```

AN587

```
00173                  MOVLW   0x080           ; 4-bit high nibble xfer
00174          endif
00175                  MOVWF   LCD_DATA        ;
00176                  BSF     LCD_CNTL, E     ;
00177                  BCF     LCD_CNTL, E     ;
00178          endif
00179  ;
00180  ; Busy Flag should be valid after this point
00181  ;
00182                  MOVLW   DISP_ON         ;
00183                  CALL    SEND_CMD        ;
00184                  MOVLW   CLR_DISP        ;
00185                  CALL    SEND_CMD        ;
00186                  MOVLW   ENTRY_INC       ;
00187                  CALL    SEND_CMD        ;
00188                  MOVLW   DD_RAM_ADDR     ;
00189                  CALL    SEND_CMD        ;
00190  ;
00191          page
00192  ;
00193  ;Send a message the hard way
00194  0039 304D       movlw   'M'
00195  003A 205F       call    SEND_CHAR
00196  003B 3069       movlw   'i'
00197  003C 205F       call    SEND_CHAR
00198  003D 3063       movlw   'c'
00199  003E 205F       call    SEND_CHAR
00200  003F 3072       movlw   'r'
00201  0040 205F       call    SEND_CHAR
00202  0041 306F       movlw   'o'
00203  0042 205F       call    SEND_CHAR
00204  0043 3063       movlw   'c'
00205  0044 205F       call    SEND_CHAR
00206  0045 3068       movlw   'h'
00207  0046 205F       call    SEND_CHAR
00208  0047 3069       movlw   'i'
00209  0048 205F       call    SEND_CHAR
00210  0049 3070       movlw   'p'
00211  004A 205F       call    SEND_CHAR
00212
00213  004B 30C0       movlw   B'11000000'     ;Address DDRam first character, second line
00214  004C 2068       call    SEND_CMD
00215
00216  ;Demonstration of the use of a table to output a message
00217  004D 3000       movlw   0               ;Table address of start of message
00218  004E 00B0  dispmsg  movwf   TEMP1        ;TEMP1 holds start of message address
00219
```

AN587

```
004F 2083   00220         call   Table
0050 39FF   00221         andlw  0FFh         ;Check if at end of message (zero
0051 1903   00222         btfsc  STATUS,Z     ;returned at end)
0052 2857   00223         goto   out
0053 205F   00224         call   SEND_CHAR    ;Display character
0054 0830   00225         movf   TEMP1,w
0055 3E01   00226         addlw  1            ;Point to next character
0056 284E   00227         goto   dispmsg
0057        00228 out
0057        00229 loop
0057 2857   00230         goto   loop         ;Stay here forever
            00231 ;
            00232 ;
            00233 INIT_DISPLAY
0058 300C   00234         MOVLW  DISP_ON      ; Display On, Cursor On
            00235         CALL   SEND_CMD     ; Send This command to the Display Module
0059 2068   00236         MOVLW  CLR_DISP     ; Clear the Display
005A 3001   00237         CALL   SEND_CMD     ; Send This command to the Display Module
005B 2068   00238         MOVLW  ENTRY_INC    ; Set Entry Mode Inc., No shift
005C 3006   00239         CALL   SEND_CMD     ; Send This command to the Display Module
005D 2068   00240         RETURN
005E 0008   00241 ;
            00242         page
            00243 ;************************************************************
            00244 ;* The LCD Module Subroutines                               *
            00245 ;*                                                          *
            00246 ;************************************************************
            00247 ;
            00248         ; if ( Four_bit )    ; 4-bit Data transfers?
            00249         ;   if ( Data_HI )   ; 4-bit transfers on the high nibble of the PORT
            00250 ;
            00251 ;************************************************************
            00252 ;**SendChar - Sends character to LCD                        *
            00253 ;*This routine splits the character into the upper and lower *
            00254 ;*nibbles and sends them to the LCD, upper nibble first.     *
            00255 ;************************************************************
            00256 ;
            00257 SEND_CHAR
            00258         MOVWF  CHAR         ;Character to be sent is in W
            00259         CALL   BUSY_CHECK   ;Wait for LCD to be ready
            00260         MOVF   CHAR, w
            00261         ANDLW  0xF0         ;Get upper nibble
            00262         MOVWF  LCD_DATA     ;Send data to LCD
            00263         BCF    LCD_CNTL, RW ;Set LCD to read
            00264         BSF    LCD_CNTL, RS ;Set LCD to data mode
            00265         BSF    LCD_CNTL, E  ;toggle E for LCD
            00266
```

AN587

```
00267              BCF     LCD_CNTL, E
00268              SWAPF   CHAR, w          ; Get lower nibble
00269              ANDLW   0xF0
00270              MOVWF   LCD_DATA         ; Send data to LCD
00271              BSF     LCD_CNTL, E      ; toggle E for LCD
00272              BCF     LCD_CNTL, E
00273              RETURN
00274      ;
00275              else                     ; 4-bit transfers on the low nibble of the PORT
00276      ;
00277      ;****************************************************************
00278      ;* SEND_CHAR - Sends character to LCD                          *
00279      ;* This routine splits the character into the upper and lower  *
00280      ;* nibbles and sends them to the LCD, upper nibble first.      *
00281      ;* The data is transmitted on the PORT<3:0> pins               *
00282      ;****************************************************************
00283      ;
00284      SEND_CHAR
00285              MOVWF   CHAR             ; Character to be sent is in W
00286              CALL    BUSY_CHECK       ; Wait for LCD to be ready
00287              SWAPF   CHAR, W
00288              ANDLW   0x0F             ; Get upper nibble
00289              MOVWF   LCD_DATA         ; Send data to LCD
00290              BCF     LCD_CNTL, RW     ; Set LCD to read
00291              BSF     LCD_CNTL, RS     ; Set LCD to data mode
00292              BSF     LCD_CNTL, E      ; toggle E for LCD
00293              BCF     LCD_CNTL, E
00294              MOVF    CHAR, W
00295              ANDLW   0x0F             ; Get lower nibble
00296              MOVWF   LCD_DATA         ; Send data to LCD
00297              BSF     LCD_CNTL, E      ; toggle E for LCD
00298              BCF     LCD_CNTL, E
00299              RETURN
00300      ;
00301              else
00302      ;
00303              endif
00304      ;****************************************************************
00305      ;* SEND_CHAR - Sends character contained in register W to LCD  *
00306      ;* This routine sends the entire character to the PORT         *
00307      ;* The data is transmitted on the PORT<7:0> pins               *
00308      ;****************************************************************
00309      ;
00310      SEND_CHAR
00311              MOVWF   CHAR             ; Character to be sent is in W
00312              CALL    BUSY_CHECK       ; Wait for LCD to be ready
00313              MOVF    CHAR, w

005F 00B6
005F 2071
0060 0836
0061
```

AN587

```
0062 0088
0063 1105
0064 1485
0065 1585
0067 0008

00314              MOVWF   LCD_DATA        ; send data to LCD
00315              BCF     LCD_CNTL, RW    ; Set LCD in read mode
00316              BSF     LCD_CNTL, RS    ; Set LCD in data mode
00317              BSF     LCD_CNTL, E     ; toggle E for LCD
00318              BCF     LCD_CNTL, E
00319              RETURN
00320      ;
00321      endif
00322      ;
00323      page
00324 ;*********************************************************
00325 ;* SendCmd - Sends command to LCD                        *
00326 ;* This routine splits the command into the upper and lower *
00327 ;* nibbles and sends them to the LCD, upper nibble first. *
00328 ;* The data is transmitted on the PORT<3:0> pins          *
00329 ;*********************************************************
00330 ;
00331      if ( Four_bit )               ; 4-bit Data transfers?
00332 ;
00333          if ( Data_HI )            ; 4-bit transfers on the high nibble of the PORT
00334 ;
00335 ;*********************************************************
00336 ;* SEND_CMD - Sends command to LCD                       *
00337 ;* This routine splits the command into the upper and lower *
00338 ;* nibbles and sends them to the LCD, upper nibble first. *
00339 ;*********************************************************
00340 ;
00341 ;
00342 SEND_CMD     MOVWF   CHAR            ; Character to be sent is in W
00343              CALL    BUSY_CHECK      ; Wait for LCD to be ready
00344              MOVF    CHAR, w
00345              ANDLW   0xF0            ; Get upper nibble
00346              MOVWF   LCD_DATA        ; Send data to LCD
00347              BCF     LCD_CNTL, RW    ; Set LCD to read
00348              BCF     LCD_CNTL, RS    ; Set LCD to command mode
00349              BSF     LCD_CNTL, E     ; toggle E for LCD
00350              BCF     LCD_CNTL, E
00351              SWAPF   CHAR, w
00352              ANDLW   0xF0            ; Get lower nibble
00353              MOVWF   LCD_DATA        ; Send data to LCD
00354              BSF     LCD_CNTL, E     ; toggle E for LCD
00355              BCF     LCD_CNTL, E
00356              RETURN
00357 ;
00358          else                       ; 4-bit transfers on the low nibble of the PORT
00359 ;
00360 ;
```

AN587

```
00361 SEND_CMD    MOVWF   CHAR            ; Character to be sent is in W
00362             CALL    BUSY_CHECK      ; Wait for LCD to be ready
00363             SWAPF   CHAR, W
00364             ANDLW   0x0F            ; Get upper nibble
00365             MOVWF   LCD_DATA        ; Send data to LCD
00366             BCF     LCD_CNTL, RW    ; Set LCD to read
00367             BCF     LCD_CNTL, RS    ; Set LCD to command mode
00368             BSF     LCD_CNTL, E     ; toggle E for LCD
00369             BCF     LCD_CNTL, E
00370             MOVF    CHAR, W         ; Get lower nibble
00371             ANDLW   0x0F
00372             MOVWF   LCD_DATA        ; Send data to LCD
00373             BSF     LCD_CNTL, E     ; toggle E for LCD
00374             BCF     LCD_CNTL, E
00375             RETURN
00376 ;
00377             endif
00378       else
00379 ;
00380 ;**************************************************************
00381 ;* SEND_CND - Sends command contained in register W to LCD   *
00382 ;* This routine sends the entire character to the PORT       *
00383 ;* The data is transmitted on the PORT<7:0> pins             *
00384 ;**************************************************************
00385 ;
00386 ;
0068  00B6   00387 SEND_CMD    MOVWF   CHAR            ; Command to be sent is in W
0069  2071   00388             CALL    BUSY_CHECK      ; Wait for LCD to be ready
006A  0836   00389             MOVF    CHAR, W
006B  0088   00390             MOVWF   LCD_DATA        ; send data to LCD
006C  1105   00391             BCF     LCD_CNTL, RW    ; Set LCD in read mode
006D  1085   00392             BCF     LCD_CNTL, RS    ; Set LCD in command mode
006E  1585   00393             BSF     LCD_CNTL, E     ; toggle E for LCD
006F  1185   00394             BCF     LCD_CNTL, E
0070  0008   00395             RETURN
00396 ;
00397             endif
00398       page
00399 ;
00400       if ( Four_bit )                            ; 4-bit Data transfers?
00401 ;
00402         if ( Data_HI )                           ; 4-bit transfers on the high nibble of the PORT
00403 ;
00404 ;**************************************************************
00405 ;**************************************************************
00406 ;* This routine checks the busy flag, returns when not busy   *
00407 ;**************************************************************
```

AN587

```
00408 ;* Affects:                                                        *
00409 ;*    TEMP - Returned with busy/address                            *
00410 ;******************************************************************
00411 ;
00412 BUSY_CHECK
00413         BSF     STATUS, RP0        ; Select Register Bank1
00414         MOVLW   0xFF
00415         MOVWF   LCD_DATA_TRIS      ; Set Port_D for input
00416         BCF     STATUS, RP0        ; Select Register Bank0
00417         BCF     LCD_CNTL, RS       ; Set LCD for Command mode
00418         BSF     LCD_CNTL, RW       ; Setup to read busy flag
00419         BSF     LCD_CNTL, E        ; Set E high
00420         BCF     LCD_CNTL, E        ; Set E low
00421         MOVF    LCD_DATA, W        ; Read upper nibble busy flag, DDRam address
00422         ANDLW   0xF0               ; Mask out lower nibble
00423         MOVWF   TEMP
00424         BSF     LCD_CNTL, E        ; Toggle E to get lower nibble
00425         BCF     LCD_CNTL, E
00426         SWAPF   LCD_DATA, w        ; Read lower nibble busy flag, DDRam address
00427         ANDLW   0x0F               ; Mask out upper nibble
00428         IORWF   TEMP, F            ; Combine nibbles
00429         BTFSC   TEMP, 7            ; Check busy flag, high = busy
00430         GOTO    BUSY_CHECK         ; If busy, check again
00431         BCF     LCD_CNTL, RW
00432         BSF     STATUS, RP0        ; Select Register Bank1
00433         MOVLW   0x0F
00434         MOVWF   LCD_DATA_TRIS      ; Set Port_D for output
00435         BCF     STATUS, RP0        ; Select Register Bank0
00436         RETURN
00437 ;
00438         else                       ; 4-bit transfers on the low nibble of the PORT
00439 ;******************************************************************
00440 ;* This routine checks the busy flag, returns when not busy       *
00441 ;* Affects:                                                        *
00442 ;*    TEMP - Returned with busy/address                            *
00443 ;******************************************************************
00444 ;
00445 ;
00446 BUSY_CHECK
00447         BSF     STATUS, RP0        ; Bank 1
00448         MOVLW   0xFF
00449         MOVWF   LCD_DATA_TRIS      ; Set PortB for input
00450         BCF     STATUS, RP0        ; Bank 0
00451         BCF     LCD_CNTL, RS       ; Set LCD for Command mode
00452         BSF     LCD_CNTL, RW       ; Setup to read busy flag
00453         BSF     LCD_CNTL, E        ; Set E high
00454         BCF     LCD_CNTL, E        ; Set E low
```

AN587

```
00455               SWAPF   LCD_DATA, W      ; Read upper nibble busy flag, DDRam address
00456               ANDLW   0xF0             ; Mask out lower nibble
00457               MOVWF   TEMP
00458               BSF     LCD_CNTL, E      ; Toggle E to get lower nibble
00459               BCF     LCD_CNTL, E
00460               MOVF    LCD_DATA, W      ; Read lower nibble busy flag, DDRam address
00461               ANDLW   0x0F             ; Mask out upper nibble
00462               IORWF   TEMP, F          ; Combine nibbles
00463               BTFSC   TEMP, 7          ; Check busy flag, high = busy
00464               GOTO    BUSY_CHECK       ; If busy, check again
00465               BCF     LCD_CNTL, RW
00466               BSF     STATUS, RP0      ; Bank 1
00467               MOVLW   0xF0
00468               MOVWF   LCD_DATA_TRIS    ; RB7 - 4 = inputs, RB3 - 0 = output
00469               BCF     STATUS, RP0      ; Bank 0
00470               RETURN
00471  ;
00472          else
00473  ;
00474  ;********************************************************
00475  ;* This routine checks the busy flag, returns when not busy  *
00476  ;*                                                          *
00477  ;* Affects:                                                 *
00478  ;*     TEMP - Returned with busy/address                    *
00479  ;********************************************************
00480  ;
00481  BUSY_CHECK
00482               BSF     STATUS, RP0      ; Select Register Bank1
00483               MOVLW   0xFF
00484               MOVWF   LCD_DATA_TRIS    ; Set port_D for input
00485               BCF     STATUS, RP0      ; Select Register Bank0
00486               BCF     LCD_CNTL, RS     ; Set LCD for command mode
00487               BSF     LCD_CNTL, RW     ; Setup to read busy flag
00488               BSF     LCD_CNTL, E      ; Set E high
00489               BCF     LCD_CNTL, E      ; Set E low
00490               MOVF    LCD_DATA, w      ; Read busy flag, DDRam address
00491               MOVWF   TEMP
00492               BTFSC   TEMP, 7          ; Check busy flag, high=busy
00493               GOTO    BUSY_CHECK
00494               BCF     LCD_CNTL, RW
00495               BSF     STATUS, RP0      ; Select Register Bank1
00496               MOVLW   0x00
00497               MOVWF   LCD_DATA_TRIS    ; Set port_D for output
00498               BCF     STATUS, RP0      ; Select Register Bank0
00499               RETURN
00500  ;
00501          endif
```

```
0071 1683
0072 30FF
0073 0088
0074 1283
0075 1085
0076 1505
0077 1585
0078 1185
0079 0808
007A 00B5
007B 1BB5
007C 2871
007D 1105
007E 1683
007F 3000
0080 0088
0081 1283
0082 0008
```

AN587

```
                       00502         page
                       00503  ; Table
0083 0782              00504  Table       addwf   PCL, F    ; Jump to char pointed to in W reg
0084 344D              00505              retlw   'M'
0085 3469              00506              retlw   'i'
0086 3463              00507              retlw   'c'
0087 3472              00508              retlw   'r'
0088 346F              00509              retlw   'o'
0089 3463              00510              retlw   'c'
008A 3468              00511              retlw   'h'
008B 3469              00512              retlw   'i'
008C 3470              00513              retlw   'p'
008D 3420              00514              retlw   ' '
008E 3454              00515              retlw   'T'
008F 3465              00516              retlw   'e'
0090 3463              00517              retlw   'c'
0091 3468              00518              retlw   'h'
0092 346E              00519              retlw   'n'
0093 346F              00520              retlw   'o'
0094 346C              00521              retlw   'l'
0095 346F              00522              retlw   'o'
0096 3467              00523              retlw   'g'
0097 3479              00524              retlw   'y'
0098                   00525
0098 3400              00526  Table_End   retlw   0
                       00527  ;
                       00528          if ( (Table & 0x0FF) >= (Table_End & 0x0FF) )
                       00529          MESSG   "Warning - User Definded: Table Table crosses page boundry in computed jump"
                       00530          endif
                       00531  ;
                       00532          ;
                       00533
                       00534
                       00535          end
                       00536
```

AN587

```
MEMORY USAGE MAP ('X' = Used,  '-' = Unused)

0000 : X--XXXXXXXXXXX XXXXXXXXXXXXXXXX XXXXXXXXXXXXXXXX XXXXXXXXXXXXXXXX
0040 : XXXXXXXXXXXXXXXX XXXXXXXXXXXXXXXX XXXXXXXXXXXXXXXX XXXXXXXXXXXXXXXX
0080 : XXXXXXXXXXXXXXXX XXXXXXXXX------- ---------------- ----------------

All other memory blocks unused.

Program Memory Words Used:   150
Program Memory Words Free:  1898

Errors   :   0
Warnings :   0 reported,     0 suppressed
Messages :   0 reported,    12 suppressed
```

CHAPTER 9

A HANDY-DANDY LOW-POWER, LOW-COST DIGITAL CLOCK

INTRODUCTION

This project is just plain fun! Not only does it produce a useful digital clock (see *Figure 9.3*), it also illustrates some interesting and valuable features of the PIC16C54, as well as the PICMICRO® MCU line in general. Hence, I found this one—which is Microchip application note AN615 (*Figure 9.2*), authored by John Day of Field Applications Engineering—a natural for this book. It fits perfectly into our goal of learning to use PICMICRO® microcontrollers.

The first thing you will notice about the circuit in *Figure 9.1* is its simplicity. I have built digital clocks in the past from dedicated integrated circuits, and I was amazed at the difference in the configuration of those projects as compared to this one. Using the PIC16C54, the circuitry is nothing short of "bare bones." The secret here is, of course, the software, and I will discuss that at length.

But, not only is the layout minimal, it is extremely clever in its nature—clever and low in cost! For example, the

I have built digital clocks in the past from dedicated integrated circuits, and I was amazed at the difference in the configuration of those projects as compared to this one.

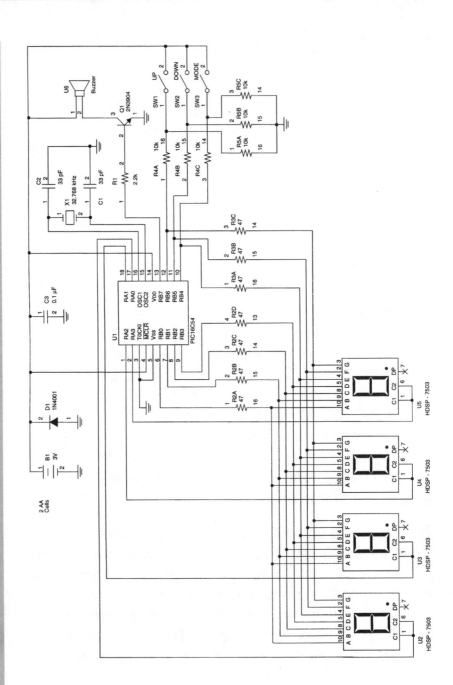

**Figure 9.1.
Schematic
diagram of the
low-power, low-
cost digital
clock.**

PIC16C54 is the least expensive device of the Microchip line, and very power efficient. By applying the attributes of the PIC16C54 and some other modern components, you will end up with a full-function digital alarm clock/count-down timer that runs on just two "AA" batteries. And that includes a more visible light-emitting diode (LED) display.

As a result, I predict (just call me the amazing Kreskin) that you will like this little gem. It's simple to construct, easy on your pocketbook, and you get a very functional digital clock in the bargain. Sally bar the door, you can't beat that deal!

HARDWARE

Referring to Figure 9.1, the small component count is immediately obvious. You can almost count all the parts on both hands. That's if you are into counting components on your fingers. The heart of this system is, of course, the PIC16C54 microcontroller. The PIC16C54 has 12 input/output (I/O) lines available, and this design uses all of them. Similar to the voltmeter in Chapter 5, the RB0 to RB6 lines handle the segments (a, b, c, d, e, f, and g) of a four-digit multiplexed light-emitting diode (LED) display, and RA0 to RA3 provide the current sink for the individual display modules.

Seven 47-ohm dropping resistors (R2A-R2D and R3A-R3C) are employed on the segment driver lines to accommodate the LED voltage characteristics, and this could be a "resistor network" to further simplify construction. With the assistance of transistor Q1, a 2N3904 or equivalent, the last RB line (RB7) is used to activate the "buzzer" for the alarm function.

Employing CMOS architecture, the microcontroller consumes a very small amount of power, making it ideal for battery operation.

Also connected to lines RB4, RB5, and RB6 are the three control switches used to set time and mode. Again, buffer resistors are employed (R4A-R4C) and these are 10-kilohm units. The other side of each switch is tied to the positive rail, as is the unconnected side of the buzzer.

Last, we have the clocking circuit that consists of a quartz crystal and two capacitors. Since the PIC16C54 is compatible with those little 32.768 kilohertz watch crystals, that is what was selected for this project. They are small and *cheap*! The crystal is wired across the PIC16C54's oscillator lines (OSC1 and OSC2), with the two 33 picofarad capacitors (C1 and C2) connected in series with the lines to ground. These are incorporated to help "kick off" the oscillator, and may not be needed. You will have to try the circuit without them to determine if they are necessary.

Power is provided to the system in the form of two 1½-volt "AA" alkaline batteries. This source is filtered by a 0.1-microfarad capacitor (C3), and a 1N4001 diode (D1) is used to protect the microcontroller against accidental reverse battery polarity. Naturally, the negative side of the supply is connected to the ground rail, while the positive side connects to the various places where a positive potential is needed (positive rail). I'm not trying to insult anyone's intelligence with that last statement; just trying to be thorough.

That is pretty much the "nuts and bolts" of the hardware design. I can't help but admire a circuit with this degree of design efficiency. But, before we move on, let me cover the hardware scheme in a little more detail. In doing so,

I think you will pick up a few tips on how to prototype similar projects in the future.

For starters, let's look at how the design takes advantage of the PIC16C54's attributes. Employing complementary metal oxide semiconductor (CMOS) architecture, the microcontroller consumes a very small amount of power, making it ideal for battery operation. Like the PIC16C71, each port line will sink about 25 milliamps and source around 20 milliamps, and that is perfect for directly driving an LED display.

As for the display, the app notes recommend four high-efficiency HDSP-7503 common cathode LED modules (U2 to U5) that draw a very conservative 3.5 milliamps of current. This decision helps with the minimal component count by eliminating the need for external driver transistors.

Additionally, since the PIC16C54 sinks more current than it sources, common cathode devices are a natural choice. Regarding the piezo buzzer, it is of the low-impedance, direct-drive persuasion, further reducing the power load. By the way, the buzzer tone is determined by the software, and that will be covered later.

It goes without saying that the life of the battery is directly related to the load on said battery. One way to improve the strain on a battery supply is to operate the system at a low frequency. Since the PIC16C54—using a very efficient instruction set—is able to execute program instruction in one cycle, the 32.768-kilohertz crystal easily fits the bill for this design. Hence, this reduced frequency aids the circuit in being power stingy.

As a short sidebar, let me elaborate on the clocking system. The PIC16C54 input clock has four internal phases that create an internal instruction cycle. The formula, 1/(CLKIN/4), determines the instruction time, with "CLKIN" representing the actual crystal frequency. Thus, this system has an instruction time of 122.07 microseconds as per the above formula (1/(32.768/4) = 1/8.192). This means that each instruction executes in 122.07 microseconds, for 8,192 cycles per second. By comparison, this is *slow*, but that speed does enhance the power-conservative nature of the system.

As a final hardware note, let's discuss the display and control switches. In review, there are four seven-segment common cathode display units arranged as in a multiplexed configuration, and three selector switches incorporated into this design. Multiplexing is chosen for both power conservation and to accommodate the number of port lines available on the PIC16C54. As with Chapter 5's digital voltmeter, the limited number of input/output (I/O) ports dictates multiplexing. In this case, individual segment addressing would require 28 lines, and that doesn't include the line needed for the buzzer.

Regarding the low-power operation feature of this project, multiplexing again helps. In such a scheme, only one display is active, or "on", at any given moment, and that means the display is using just one-fourth of the power it would use if all four displays were illuminated. So, this is usually a consideration in any design using an LED display. It also makes things simple.

The circuit uses 11 of the 12 available I/O lines to light the display, with the four RA ports handling the common cathodes and seven of the eight RB lines managing the separate segment (a, b, c, d, e, f, and g). This is done partially to allow for "moves" and "rotates." The last RB port (RB7) is utilized for the buzzer.

The three switches (keys) are present for time setting, and these are connected (multiplexed) to three of the RB lines (RB4, RB5, and RB6). Again, this is done to oblige the small number of overall I/O ports.

With that said, you should have a solid idea of how the hardware functions. As I mentioned earlier, the circuitry is fundamental in nature, and most of the work is done by the software. So, let's have a look at that code, and how it does its job.

SOFTWARE SCHEME

For the clock to function properly, the main software loop must accomplish five separate chores. They are as follows.

1) A one-second elapsed time period must be detected as a result of four changes in the state of bit 7 on the TMRO register. The time display must then be incremented by 1, or the system must decrement the countdown timer by 1.
2) If either the time alarm or the countdown alarm should be ringing, or are ringing, as determined by the program, then the buzzer is activated.
3) Activation of any of the switches must be detected. If the MODE switch is pressed, then MODE is incremented. The display will be changed appropriately if either the UP switch or DOWN switch is activated.

4) To conserve power, the program must automatically turn the display "on" and "off" as dictated by the need for time display.
5) The display must be multiplexed every 32 instruction cycles (3.9 milliseconds).

If all of this is accomplished, the software is performing correctly. The clock will, in turn, keep perfect time for you or act as a countdown timer that monitors a specific interval.

As with most computer systems, a number of general-purpose registers are employed to get the job done. As you probably know, registers are areas within the microcontroller where small amounts of data can be stored or transferred from. They are quite handy, so let's look at the registers for our digital clock.

1) DISPSEGS_A to DISPSEGS_D: Standing for "display segments A to D", these are used to store the BDC information for the individual LED display modules.

2) CLK_SEC: Standing for "clock seconds", this stores the seconds count.

3) CLK_MIN_LD, CLK_MIN_HD: Standing for "clock minute low digit, clock minute high digit", these store the current time upper and lower minute digits.

4) CLK_HOUR_LD, CLK_HOUR_HD: Standing for "clock hour low digit, clock hour high digit", these store the current time upper and lower hour digits.

5) ALM_MIN_LD, ALM_MIN_HD: Standing for "alarm minute low digit, alarm minute high digit", these store the alarm time upper and lower minute digits.

6) ALM_HOUR_LD, ALM_HOUR_HD: Standing for "alarm hour low digit, alarm hour high digit", these store the alarm time upper and lower hour digits.

7) TMR_SEC_LD, TMR_SEC_HD: Standing for "timer second low digit, timer second high digit", these store the timer upper and lower second digits.

8) TMR_MIN_LD, TMR_MIN_HD: Standing for "timer minute low digit, timer minute high digit", these store the timer upper and lower minute digits.

9) KEYPAT: Standing for "key pattern", this stores the pattern of the keys currently activated: UP = bit6, DOWN = bit5, and MODE = bit4.

10) FLAGS: Standing for "flags" (Ha, Ha), these store the present flag bits that control such things as the alarm "on", mode, etc.

11) PREVTMRO: Standing for "previous TMRO", this is used to compare two TMRO values by storing the previous TMRO value.

12) TEMP: Standing for "temporary", this is a multipurpose temporary register.

13) DISCONCNT: Standing for "display "on" counter", this stores the remaining seconds the display should remain "on".

14) MODE_COUNT: Standing for "mode counter", this stores the length of time, in half-second intervals, the MODE, UP or DOWN switches have been activated. Primarily used to change from minute to hour settings.

15) ALARMCNT: Standing for "alarm counter", this stores the number of remaining beeps to be sent to the buzzer.

As with most computer systems, a number of general-purpose registers are employed to get the job done.

Each register is set up to do a specific job, thus helping the program to function flawlessly.

These are fairly self-explanatory. Each register is set up to do a specific job, thus helping the program to function flawlessly. However, the FLAGS register does require a little more explanation. In an effort to determine the modes or state of the code routine, most designs need a flag or "state" bit to identify these parameters. The following describes the FLAGS register for the digital clock.

1) Bits 0,1 are used to indicate the operational mode as dictated by pressing the MODE switch. For example, 00 = Display Off, 01 = Display/Set countdown timer, 10 = Display/Set Alarm Time and 11 = Display/Set Clock.
2) Bit 3 denotes if the alarm time is equal to the current time.
3) Bit 4 indicates if the countdown timer is equal to zero.
4) Bits 5,6 handle keeping track of the seconds through a "divide by 4" counter scheme.

As can be seen, these registers are designed to manage all operations of the clock/timer. Working in conjunction with the software routines, the registers provide the necessary program maintenance. And, speaking of the software routines, let's take a look at these next. The app notes break the routines down into the following categories, so we will examine them in that fashion. Each routine will be accompanied by a short explanation of how the code works.

1) BUZZ_NOW ROUTINE: Outputs the buzzing tone during alarm for 156 ms.

This is a nifty little routine that, in the spirit of multitasking, mutliplexes two functions in order to do both at seemingly the same time. Those two functions are keeping time and buzzing the buzzer. As stated in the hardware section, this circuit uses a low-impedance, directly driven piezo buzzer that is software controlled. That is, both the pitch and when it sounds are determined by the code. The "buzz" routine is called by the main software loop and activates the buzzer for 156 milliseconds at a time at a frequency of 1,638 hertz. The routine first clears PORTB (turns "off" the display) then employs the TEMP register to count 256 pulses. Each pulse constitutes one 156-ms signal to the buzzer, and these are accomplished by the "BSF BUZZEROUT" and "BCF BUZZEROUT" instructions. Finally, after 256 pulses, the routine returns.

2) TASK_SCAN ROUTINE: Multiplex LEDs to display the next digit, only one digit is lit at a time.

Since the PIC16C54 does not have a hardware interrupt, this routine is "called" within the multiplexing time frame of every 32 instruction cycles, or 3.9 milliseconds. This is done to maintain proper display performance and must be done after every 11 instruction cycles, as the routine takes 21 cycles to complete. First, the routine aligns itself with the TMRO to ensure that, regardless of when the routine is called, scanning occurs at the same point in time. PREVSCAN is now rotated setting up the CARRY bit. The BCD code for the next digit to be displayed is moved to the W register, and the display is blanked. To select the next digit, PORTA is rotated, and the BCD code is moved to PORTB and displayed. Last, PORTA is restored from the "PREVSCAN" state.

3) DISP_VALUE ROUTINE: Update the display registers with the bitmap of what digits are to be displayed next.

The W register holds, in register file, the base address for all four displays that are to be displayed. Hence, the W register is moved to the FRS register so the indirect address register contains the first digit to be displayed. This is done because this routine, in an effort to reduce the code count and simplify the process, uses indirect addressing. Now, the first digit is converted to BCD code (segment bit pattern), then moved to DISPSEGS_A. The FSR is then incremented, and the process starts all over again. TASK_SCAN is called throughout this routine to preserve proper multiplexing.

4) TURNON_SCAN ROUTINE: Turns on the LEDs and restores a legal scan position.

In the essence of battery conservation, the clock display is turned off after eight seconds if no switch action is detected. Here, the DISPON bit is employed to pre-set the "on" display time to eight seconds. This routine sets the flag to later turn the display "on", and then checks PREVSCAN for a legal value. If the displays are "off", it restores a legal value.

5) SCAN_KEYS ROUTINE: Turns off LEDs for a moment and scans the pushbutton inputs.

Multiplexing of the control keys to the PORTB lines allows the number of I/O lines to be sufficiently reduced. The first step is to clear PORTB and set PORTA to "0Fh". That turns the display off. Now, bits 4, 5, and 6 of PORTB are set up as input to detect switch action. TEMP is loaded with the keys that have changed state, and a "0 = not pressed, 1 = pressed" pattern is placed in KEYPAT to handle switches that are hit. Last, PORTA is returned to scan and PORTB is restored to outputs.

6) CHECK_TIME ROUTINE: Checks for alarm or countdown timer expiration.

In the main program, the buzzer is activated by flag bits EGGNOW and ALARMNOW, and the process starts by setting both. This is because each second and alarm condition must be detected, and the buzzer sounded if the alarm is set. Through subtraction and a test of the STATUS register Z bit, the current time is compared with the alarm time. If these don't equate, the ALARMNOW bit is cleared. Last, the countdown seconds and minutes are compared to zero, and a non-equate results in EGGNOW being cleared. Here again, the TASK_SCAN routine is used to maintain proper multiplexing.

7) INC_TIME ROUTINE: Adds one second, minute, or hour to the clock alarm or timer.

Here is another application of "indirect addressing" for code conservation and simplification of operation. This routine is called once each second by the main program to increment the clock. Also, it will be called when the MODE key is held down and the UP key is used to modify the current or alarm time setting. Before the routine is called, however, the W register is loaded with the clock second register address. The routine is now run and the FSR register receives the value, incrementing the indirect address register. With the second counter incremented, a check for "overflow" is run and if there is none, the routine returns. If there is overflow, the minute low digit is incremented. Again, an overflow test in conducted, and the process continues until the count is correctly displayed. Additionally, this routine can be called to set the clock or alarm with the INC_MIN_LD, which only brings up the minutes. In the same fashion, only hours can be brought up with INC_HOUR_LD. Finally, the FLAGS register (bits 0 and 1) is employed to determine the mode and ensure proper overflow calculations. Naturally, to preserve multiplexing, the TASK_SCAN register is called. That poor thing is really overworked!

8) DEC_TIME ROUTINE: Subtract one second, minute, or hour from the clock alarm or timer.

This one is used for the countdown timer, and is called by the main loop whenever the timer is in use. Also, it is called when the DOWN key is pressed and the MODE is held to set current time, alarm time, or the timer. As with the preceding routine, indirect addressing is employed and for the same reasons. Again the W register is loaded with the countdown timer's second register address, and

In the main program, the buzzer is activated by flag bits EGGNOW and ALARMNOW, and the process starts by setting both. This is because each second and alarm condition must be detected.

then the routine is called. Next, the FSR register receives this value, and the indirect address register is incremented (this routine is very similar to the last one). Following decrementing of the seconds counter LSD, the overflow condition is tested and if valid, the minutes counter MSD is decremented, and the seconds counter is set to 9. Hours can be controlled in the same fashion by calling DEC_HOUR_LD_VEC, and the FLAGS register (bits 0 and 1) is employed to maintain proper overflow calculation. TASK_SCAN ensures proper multiplexing.

9) MAIN_LOOP ROUTINE: Calls the above routines as needed and keeps track of when to increment the clock or decrement the countdown timer.

This is the "Big Kahuna!" All of the above-mentioned routines are controlled (called) by this one—the main loop. And it is this code that keeps the clock "clocking." To begin, the OPTION register is loaded with the value of 03h which sets the prescaler to a "divide by 16" status. This, in turn, regulates the TMR0 and incrementation of the internal instruction cycle. Since the instruction cycle is 122.07 microseconds (remember the sidebar on the clocking system), bit0 will change every 122.07 x 16 or 1.953 milliseconds. On the other hand, bit7 will change every 122.07 x 16 x 128, or 250 milliseconds, and FLAGS register bits 5 and 6 divide this 250 milliseconds by 4 to call INC_TIME once each second. Following INC_TIME, the CHECK_TIME routine is called, and this sets the EGGNOW and ALARMNOW flag bits. The buzzer will sound if the alarm is set, due to BUZZ_NOW, but the program *does* need to maintain the time updates to keep the clock on track. Every 500 milliseconds, the switches (keys) are sampled with the "edges" (rising/falling pulse edges) detected. If either the UP or DOWN switch is activated, the buzzer will cease (the enable bits are cleared). As stated before, the MODE switch advances the clock mode, which consists of four separate modes. They are as follows.

A. Display OFF—saves battery power; defaults to this mode if no keys are pressed for eight seconds.

B. Display or Set countdown timer (holding MODE key allows setting).

C. Display or Set Alarm time (holding MODE key allows setting).

D. Display or Set Clock time (holding MODE key allows setting).

Next, the UP and DOWN keyscans are tested, with the currently displayed mode time incremented or decremented if the UP and MODE switches are activated. If just the UP or DOWN key is pressed, the display is turned "on" but not changed. To save power, after eight seconds the display will turn "off" if no switch action occurs. This is accomplished by DISPONCNT.

If either the UP or DOWN key is held, in conjunction with the MODE key, for more than four seconds, the setting is sent from the right side (minutes/seconds) to the left side (hours/minutes) by MODE_COUNT reaching zero. The last task of the main loop is to call DISP_VALUE and check the status of DISPONCNT. If DISPONCNT is at zero, the display is turned "off".

So, as can be seen, the main loop has its work cut out for it. It is also easy to see how the entire operation of this digital clock/countdown timer relies on this one routine. The next subject is the three lookup tables used by the device. The purpose of a lookup table is to convert a number into a random access memory address or a bit pattern to be used by the display. And, our clock uses three such tables. These tables serve the functions of decoding the BCD info to seven segment patterns, checking the address of the current mode, and as a diagnostic aid to a manufacturer. The following is a list with an explanation of these lookup tables.

1) MODE_TIMER: Look up the address of the clock, alarm, or timer data storage RAM.
2) LED_LOOKUP: Lookup table contains the bitmap display pattern for displaying digits 0 to 9.
3) MFG_LED_LOOKUP: Lookup table contains the bitmap display pattern used for the manufacturing mode. Only one segment is lit at a time.

As you might imagine, lookup tables are very handy fellows. They can make your programming experience easier—hence, more pleasant—if used correctly. For more information on this subject, see Chapter 6.

On a final note, there are two miscellaneous routines that are used for manufacturing tests and initialization. So, to be complete, let's take a look at these. The first is INT, and I'll bet you're saying to yourself, "That one is for initialization." Guess what? You're right. What this does is set the default time values, initializes the RAM to "0", and sets up the I/O ports. It is as simple as that. MFG_SELFTEST, on the other hand, is a diagnostic routine that is similar to the one your personal computer (PC) runs each time you boot up. In this case, it checks the LEDs, buzzer, and keys, and lets you know if there are any problems. It is used almost exclusively in the manufacturing scenario to test the clocks as they "roll off" the assembly line.

CONCLUSION

So, ain't this a neat one? This project really showcases the versatility of the PIC16C54 microcontroller and the efficiency of the companion software. And many of the routines used here can effectively be employed in other applications. See, I'm going to teach you something whether you want me to or not! Just kidding! I know everyone reading this text has a powerful thirst for knowledge; hence, I will never have to sneak the learning in. Right? Right!

Anyway, all joking aside, this digital clock is a prime example of how microcontrollers can make things soooooo easy when we let them. In this case, not only does the software do all the work, it keeps the component count down to a point where these little clocks are affordable by everyone. And, speaking of the software, it's only 510 words and only uses 25 bytes of RAM.

So, for that next project you have in the back of your mind, or for all your projects, it makes sense to consider a PICMICRO® microcontroller approach. Not only can it save you time and expense, it will provide an interesting learning experience.

This digital clock is a prime example of how microcontrollers can make things soooooo easy when we let them.

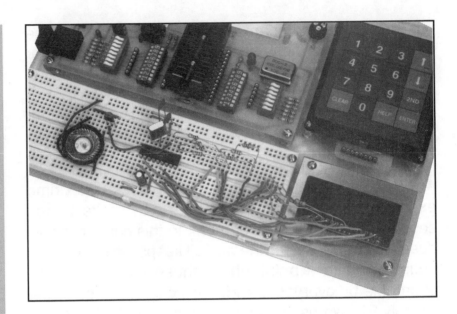

*Figure 9.3.
Here's the final
product on the
prototype
breadboard.
Note the
simplicity of the
hardware.*

AN615

APPENDIX A: CODE

MPASM 01.21.03 Intermediate   CLK8.ASM 8-21-1995 9:17:56   PAGE 1

```
LOC     OBJECT  LINE SOURCE TEXT
VALUE   CODE

                00001 ; **********************************************
                00002 ; *      PIC Egg Timer Give-Away               *
                00003 ; *                                            *
                00004 ; * Author:   John Day                         *
                00005 ; *           Sr. Field Applications Engineer  *
                00006 ; *           Northeast Region                 *
                00007 ; *                                            *
                00008 ; * Revision: 1.2                              *
                00009 ; * Date      September 22, 1994               *
                00010 ; * Part:     PIC16C54-LP/P or PIC16LC54A/P    *
                00011 ; * Fuses:    OSC:  LP                         *
                00012 ; *           WDT:  OFF                         *
                00013 ; *           Port: OFF                         *
                00014 ; *           CP:   OFF                         *
                00015 ; **********************************************
                00016 ;
                00017 ; This program is intended to run on a 32 Khz watch crystal and
                00018 ; connects to four multiplexed seven segment displays. It displays the
                00019 ; current time, alarm time and egg count down timers. There are
                00020 ; switches that allow the user to set the alarm, timer and clock functions.
                00021
                00022 LIST F=INHX8M,P=16C54
                00023 INCLUDE "p16C5X.inc"
                00001       LIST
                00002 ; P16C5X.INC  Standard Header File, Version 2.02  Microchip Technology, Inc.
                00143       LIST
OFFF    OFF8    00024       __FUSES _CP_OFF&_WDT_OFF&_LP_OSC
                00025
0007            00026       ORG 07h
                00027 ; ******************************************
                00028 ; * Static RAM Register File Definitions *
                00029 ; ******************************************
00000000        00030 INDADDR    EQU  0     ; Indirect address register
00000007        00031 DISPSEGS_A EQU  07h   ; Current Display A segment bit pattern
00000008        00032 DISPSEGS_B EQU  08h   ; Current Display B segment bit pattern
00000009        00033 DISPSEGS_C EQU  09h   ; Current Display C segment bit pattern
0000000A        00034 DISPSEGS_D EQU  0Ah   ; Current Display D segment bit pattern
0000000B        00035 CLK_SEC    EQU  0Bh   ; Clock second counter (0-59)
0000000C        00036 CLK_MIN_LD EQU  0Ch   ; Clock minute low digit counter (0-9)
0000000D        00037 CLK_MIN_HD EQU  0Dh   ; Clock minute high digit counter (0-5)
0000000E        00038 CLK_HOUR_LD EQU 0Eh   ; Clock hour low digit counter (0-9)
0000000F        00039 CLK_HOUR_HD EQU 0Fh   ; Clock hour high digit counter (0-2)
00000010        00040 ALM_MIN_LD EQU  10h   ; Alarm minute low digit counter (0-9)
00000011        00041 ALM_MIN_HD EQU  11h   ; Alarm minute high digit counter (0-5)
00000012        00042 ALM_HOUR_LD EQU 12h   ; Alarm hour lor digit counter (0-9)
00000013        00043 ALM_HOUR_HD EQU 13h   ; Alarm hour high digit counter (0-2)
00000014        00044 TMR_SEC_LD EQU  14h   ; Timer second low digit counter (0-9)
00000015        00045 TMR_SEC_HD EQU  15h   ; Timer second high digit counter (0-5)
00000016        00046 TMR_MIN_LD EQU  16h   ; Timer hour low digit counter (0-9)
00000017        00047 TMR_MIN_HD EQU  17h   ; Timer hour high digit counter (0-2)
00000018        00048 KEYPAT     EQU  18h   ; Currently pressed key bits
00000019        00049 FLAGS      EQU  19h   ; Status of alarms, display on, etc.
0000001A        00050 PREVTMR0   EQU  1Ah   ; Used to determine which TMR0 bits changed
0000001B        00051 PREVSCAN   EQU  1Bh   ; Store Common Cathode display scan state
0000001C        00052 TEMP       EQU  1Ch   ; Temporary storage
```

Figure 9.2.
(Continued on next 11 pages) **Code for the low-power, low-cost digital clock.**

AN615

```
0000001D    00053 DISPONCNT        EQU     1Dh     ; Time the displays have been on
0000001E    00054 MODE_COUNT       EQU     1Eh         ; Current mode state
0000001F    00055 ALARMCNT         EQU     1Fh         ; Time the alarm has been sounding
            00056 ; ************************************
            00057 ; * Flag and state bit definitions       *
            00058 ; ************************************
            00059 #define         SECBIT      TEMP,7      ; Bit to spawn 1/4 second count
            00060 #define         SCANBIT     TMR0,0      ; Bit to spawn display MUX
            00061 #define         MODEKEY     KEYPAT,4    ; Bit for MODEKEY pressed
            00062 #define         UPKEY       KEYPAT,6    ; Bit for UPKEY pressed
            00063 #define         DOWNKEY     KEYPAT,5    ; Bit for DOWNKEY pressed
            00064 #define         MODEKEYCHG  TEMP,4      ; Bit for delta MODEKEY
            00065 #define         TIMENOW     FLAGS,7     ; Flag to indicate 1 second passed
            00066 #define         ALARMNOW    FLAGS,3     ; Flag to indicate wakeup alarm
            00067 #define         EGGNOW      FLAGS,4     ; Flag to indicate egg timer alarm
            00068 #define         ALARMOK     STATUS,PA0  ; Flag to enable wakeup alarm
            00069 #define         EGGOK       STATUS,PA1  ; Flag to enable timer alarm
            00070 #define         BUZZEROUT   PORTB,7     ; Pin for pulsing the buzzer
            00071 #define         DISPON      DISPONCNT,4 ; Bit to turn on LED displays
            00072
            00073 ; ************************************************
            00074 ; * Various Constants used throughout the program *
            00075 ; ************************************************
0000003C    00076 SEC_MAX          EQU     .60         ; Maximum value for second counter
0000000A    00077 MIN_LD_MAX       EQU     .10         ; Maximum value for low digit of minute
00000006    00078 MIN_HD_MAX       EQU     .6          ; Maximum value for high digit of minute
00000004    00079 HOUR_LD_MAX      EQU     .4          ; Maximum value for low digit of hour
00000002    00080 HOUR_HD_MAX      EQU     .2          ; Maximum value for high digit of hour
00000003    00081 OPTION_SETUP     EQU     b'00000011' ; TMR0 - internal, /16 prescale
00000007    00082 BUZINITVAL       EQU     7           ;
00000008    00083 INIT_MODE_COUNT  EQU     8           ; Digit counts to move to hour digits
00000028    00084 ALARMCYCCNT      EQU     .40         ; Alarm for 10 seconds (ALARMCYCCNT/4)
            00085
01FF        00086          ORG     01FFh              ; The PIC5X reset vector is at end of memory
01FF        00087 reset_vector
01FF 0BA8   00088          GOTO    init               ; Jump to the initialization code
            00089
0000        00090          ORG     0
            00091 ; ************************************************
            00092 ; * Current mode look-up table            *
            00093 ; ************************************************
0000        00094 mode_timer
0000 0E03   00095          ANDLW   3                  ; Mask off upper bits just in case
0001 01E2   00096          ADDWF   PCL,F              ; Jump to one of 4 look-up entries
0002 0814   00097          RETLW   TMR_SEC_LD         ; Return the address of the 99 min timer RAM
0003 0810   00098          RETLW   ALM_MIN_LD         ; Return the address of the alarm RAM
0004 080C   00099          RETLW   CLK_MIN_LD         ; Return the address of the clock RAM
0005 080C   00100          RETLW   CLK_MIN_LD         ; Return the address of the clock RAM
            00101
            00102 ; ************************************************
            00103 ; * Buzz the buzzer for 1/8 second        *
            00104 ; ************************************************
0006        00105 buzz_now
0006 0066   00106          CLRF    PORTB              ; Shut off the segments
0007        00107 buzz_now_dispon
0007 007C   00108          CLRF    TEMP               ; Buzz for 256 pulses
0008        00109 loop_buz
0008 05E6   00110          BSF     BUZZEROUT          ; Send out pulse
0009 04E6   00111          BCF     BUZZEROUT          ; Clear out the pulse
000A 02FC   00112          DECFSZ  TEMP,F             ; Decrement counter and skip when done
000B 0A08   00113          GOTO    loop_buz           ; Go back and send another pulse
000C 0800   00114          RETLW   0                  ; We are done so come back!
            00115
            00116 ; ************************************************
            00117 ; * Mux drive the next LED display digit *
            00118 ; ************************************************
```

AN615

```
000D              00119 task_scan   ; (19 (next_scan) + 2 = 21 cycles - must be called every 11 cy)
000D 0601         00120         BTFSC   SCANBIT         ; Synch up with 3.9 mS timer bit
000E 0A0D         00121         GOTO    task_scan       ; Jump back until bit is clear
                  00122
000F              00123 next_scan   ; (15 + 2 call + 2 return = 19 cycles)
000F 035B         00124         RLF     PREVSCAN,W      ; Move to the next digit select into C
0010 073B         00125         BTFSS   PREVSCAN,1      ; 0 Check if display A was on before
0011 0209         00126         MOVF    DISPSEGS_C,W    ; Place display B value into W
0012 071B         00127         BTFSS   PREVSCAN,0      ; 1 Check if display B was on before
0013 0208         00128         MOVF    DISPSEGS_B,W    ; Place display C value into W
0014 077B         00129         BTFSS   PREVSCAN,3      ; 2 Check if display C was on before
0015 0207         00130         MOVF    DISPSEGS_A,W    ; Place display D value into W
0016 075B         00131         BTFSS   PREVSCAN,2      ; 3 Check if display D was on before
0017 020A         00132         MOVF    DISPSEGS_D,W    ; Place display A value into W
0018 0066         00133         CLRF    PORTB           ; Turn off all segments
0019 037B         00134         RLF     PREVSCAN,F      ; Move to the next digit
001A 0365         00135         RLF     PORTA,F         ; Move port to the next digit
001B 0026         00136         MOVWF   PORTB           ; Place next segment value on PORTB
001C 021B         00137         MOVF    PREVSCAN,W      ; Restore the port in case it is wrong
001D 0025         00138         MOVWF   PORTA           ; Restore the port
001E 0800         00139         RETLW   0               ; Display is updated - now return
                  00140
                  00141
                  00142 ; ***********************************************
                  00143 ; * Move new digit display info out to display *
                  00144 ; ***********************************************
001F              00145 disp_value
001F 0024         00146         MOVWF   FSR             ; Place W into FSR for indirect addressing
0020 090D         00147         CALL    task_scan       ; Scan the next LED digit.
0021 0200         00148         MOVF    INDADDR,W       ; Place display value into W
0022 0937         00149         CALL    led_lookup      ; Look up seven segment value
0023 0027         00150         MOVWF   DISPSEGS_A      ; Move value out to display register A
0024 02A4         00151         INCF    FSR,F           ; Go to next display value
0025 090D         00152         CALL    task_scan       ; Scan the next LED digit.
0026 0200         00153         MOVF    INDADDR,W       ; Place display value into W
0027 0937         00154         CALL    led_lookup      ; Look up seven segment value
0028 0028         00155         MOVWF   DISPSEGS_B      ; Move value out to display register B
0029 02A4         00156         INCF    FSR,F           ; Go to next display value
002A 090D         00157         CALL    task_scan       ; Scan the next LED digit.
002B 0200         00158         MOVF    INDADDR,W       ; Place display value into W
002C 0937         00159         CALL    led_lookup      ; Look up seven segment value
002D 0029         00160         MOVWF   DISPSEGS_C      ; Move value out to display register C
002E 02A4         00161         INCF    FSR,F           ; Go to next display value
002F 090D         00162         CALL    task_scan       ; Scan the next LED digit.
0030 0200         00163         MOVF    INDADDR,W       ; Place display value into W
0031 0643         00164         BTFSC   STATUS,Z        ; ZBLANK - Check for a zero
0032 0240         00165         COMF    INDADDR,W       ; ZBLANK - Clear digit with FF if leading 0
0033 0937         00166         CALL    led_lookup      ; Look up seven segment value
0034 002A         00167         MOVWF   DISPSEGS_D      ; Move value out to display register D
0035 090D         00168         CALL    task_scan       ; Scan the next LED digit.
0036 0800         00169         RETLW   0
                  00170
                  00171 ; ***********************************************
                  00172 ; * Convert display value into segments  *
                  00173 ; ***********************************************
0037              00174 led_lookup
0037 0E0F         00175         ANDLW   0Fh             ; Strip off upper digits
0038 01E2         00176         ADDWF   PCL,F           ; Jump into the correct location
0039 083F         00177         RETLW   b'00111111'     ; Bit pattern for a Zero
003A 0806         00178         RETLW   b'00000110'     ; Bit pattern for a One
003B 085B         00179         RETLW   b'01011011'     ; Bit pattern for a Two
003C 084F         00180         RETLW   b'01001111'     ; Bit pattern for a Three
003D 0866         00181         RETLW   b'01100110'     ; Bit pattern for a Four
003E 086D         00182         RETLW   b'01101101'     ; Bit pattern for a Five
003F 087D         00183         RETLW   b'01111101'     ; Bit pattern for a Six
0040 0807         00184         RETLW   b'00000111'     ; Bit pattern for a Seven
```

AN615

```
0041 087F    00185         RETLW   b'01111111'    ; Bit pattern for a Eight
0042 086F    00186         RETLW   b'01101111'    ; Bit pattern for a Nine
0043 0800    00187         RETLW   0              ; Turn display off - ILLEGAL VALUE
0044 0800    00188         RETLW   0              ; Turn display off - ILLEGAL VALUE
0045 0800    00189         RETLW   0              ; Turn display off - ILLEGAL VALUE
0046 0800    00190         RETLW   0              ; Turn display off - ILLEGAL VALUE
0047 0800    00191         RETLW   0              ; Turn display off - ILLEGAL VALUE
0048 0800    00192         RETLW   0              ; Turn display off - ILLEGAL VALUE
             00193
             00194 ; ********************************************************************
             00195 ; * Convert display value into single segment ON for manufacturing diags *
             00196 ; ********************************************************************
0049         00197 mfg_led_lookup
0049 0E07    00198         ANDLW   07h            ; Strip off upper digits
004A 01E2    00199         ADDWF   PCL,F          ; Jump into the correct location
004B 0801    00200         RETLW   b'00000001'    ; Bit pattern for segment A on only
004C 0802    00201         RETLW   b'00000010'    ; Bit pattern for segment B on only
004D 0804    00202         RETLW   b'00000100'    ; Bit pattern for segment C on only
004E 0808    00203         RETLW   b'00001000'    ; Bit pattern for segment D on only
004F 0810    00204         RETLW   b'00010000'    ; Bit pattern for segment E on only
0050 0820    00205         RETLW   b'00100000'    ; Bit pattern for segment F on only
0051 0840    00206         RETLW   b'01000000'    ; Bit pattern for segment G on only
0052 087F    00207         RETLW   b'01111111'    ; Bit pattern for all segments on
             00208
             00209 ; *********************************************************
             00210 ; * Wake-up and turn on the displays                      *
             00211 ; *********************************************************
0053         00212 turnon_scan
0053 059D    00213         BSF     DISPON         ; Set display ON bit
0054 0CEE    00214         MOVLW   b'11101110'    ; Place digit 0 scan pattern in W
0055 019B    00215         XORWF   PREVSCAN,W     ; See if this is the current scan
0056 0643    00216         BTFSC   STATUS,Z       ; Skip if this is not the current scan
0057 0800    00217         RETLW   0              ; Legal scan value - we are done!
0058 0CDD    00218         MOVLW   b'11011101'    ; Place digit 1 scan pattern in W
0059 019B    00219         XORWF   PREVSCAN,W     ; See if this is the current scan
005A 0643    00220         BTFSC   STATUS,Z       ; Skip if this is not the current scan
005B 0800    00221         RETLW   0              ; Legal scan value - we are done!
005C 0CBB    00222         MOVLW   b'10111011'    ; Place digit 2 scan pattern in W
005D 019B    00223         XORWF   PREVSCAN,W     ; See if this is the current scan
005E 0643    00224         BTFSC   STATUS,Z       ; Skip if this is not the current scan
005F 0800    00225         RETLW   0              ; Legal scan value - we are done!
0060 0C77    00226         MOVLW   b'01110111'    ; Place digit 3 scan pattern in W
0061 019B    00227         XORWF   PREVSCAN,W     ; See if this is the current scan
0062 0643    00228         BTFSC   STATUS,Z       ; Skip if this is not the current scan
0063 0800    00229         RETLW   0              ; Legal scan value - we are done!
0064 0CEE    00230         MOVLW   0EEh           ; Move digit 0 scan value into W
0065 003B    00231         MOVWF   PREVSCAN       ; Move it into scan pattern register
             00232
             00233 ; *****************************************
             00234 ; * Scan for pressed keys                  *
             00235 ; *****************************************
0066         00236 scan_keys
0066 0066    00237         CLRF    PORTB          ; Turn off all of the segments
0067 0CFF    00238         MOVLW   0FFh           ; Place FF into W
0068 0025    00239         MOVWF   PORTA          ; Make PORT A all ones
0069 0C70    00240         MOVLW   b'01110000'    ; Place 70 into W
006A 0006    00241         TRIS    PORTB          ; Make RB4,5,6 inputs others outputs
006B 0206    00242         MOVF    PORTB,W        ; Place keyscan value into W
006C 0198    00243         XORWF   KEYPAT,W       ; Place Delta key press into W
006D 003C    00244         MOVWF   TEMP           ; Place Delta key press into TEMP
006E 01B8    00245         XORWF   KEYPAT,F       ; Update KEYPAT reg to buttons pressed
006F 0040    00246         CLRW                   ; Place 0 into W
0070 0006    00247         TRIS    PORTB          ; Make PORT B outputs
0071 021B    00248         MOVF    PREVSCAN,W     ; Place previous scan value into W
0072 0025    00249         MOVWF   PORTA          ; Turn on the scan
0073 0800    00250         RETLW   0
```

AN615

```
                  00251 ; ****************************************
                  00252 ; * Check if alarm or timer is expired   *
                  00253 ; ****************************************
0074              00254 check_time
0074 090D         00255          CALL     task_scan        ; Scan the next LED digit.
0075 0579         00256          BSF      ALARMNOW         ; Set the alarm bit
0076 0599         00257          BSF      EGGNOW           ; Set the Egg timer alarm bit
0077 0210         00258          MOVF     ALM_MIN_LD,W     ; Place alarm minute counter into W
0078 008C         00259          SUBWF    CLK_MIN_LD,W     ; CLK_MIN_LD - W -> W
0079 0743         00260          BTFSS    STATUS,Z         ; Skip if they are equal
007A 0479         00261          BCF      ALARMNOW         ; They are not equal so clear alarm bit
007B 0211         00262          MOVF     ALM_MIN_HD,W     ; Place alarm minute counter into W
007C 008D         00263          SUBWF    CLK_MIN_HD,W     ; CLK_MIN_HD - W -> W
007D 0743         00264          BTFSS    STATUS,Z         ; Skip if they are equal
007E 0479         00265          BCF      ALARMNOW         ; They are not equal so clear alarm bit
007F 090D         00266          CALL     task_scan        ; Scan the next LED digit.
0080 0212         00267          MOVF     ALM_HOUR_LD,W    ; Place alarm hour counter into W
0081 008E         00268          SUBWF    CLK_HOUR_LD,W    ; CLK_HOUR_LD - W -> W
0082 0743         00269          BTFSS    STATUS,Z         ; Skip if they are equal
0083 0479         00270          BCF      ALARMNOW         ; They are not equal so clear alarm bit
0084 0213         00271          MOVF     ALM_HOUR_HD,W    ; Place alarm hour counter into W
0085 008F         00272          SUBWF    CLK_HOUR_HD,W    ; CLK_HOUR_LD - W -> W
0086 0743         00273          BTFSS    STATUS,Z         ; Skip if they are equal
0087 0479         00274          BCF      ALARMNOW         ; They are not equal so clear alarm bit
0088 090D         00275          CALL     task_scan        ; Scan the next LED digit.
0089 0214         00276          MOVF     TMR_SEC_LD,W     ; Set the Z bit to check for zero
008A 0743         00277          BTFSS    STATUS,Z         ; Skip if this digit is zero
008B 0499         00278          BCF      EGGNOW           ; Timer is not zero so clear egg alarm bit
008C 0215         00279          MOVF     TMR_SEC_HD,W     ; Set the Z bit to check for zero
008D 0743         00280          BTFSS    STATUS,Z         ; Skip if this digit is zero
008E 0499         00281          BCF      EGGNOW           ; Timer is not zero so clear egg alarm bit
008F 0216         00282          MOVF     TMR_MIN_LD,W     ; Set the Z bit to check for zero
0090 0743         00283          BTFSS    STATUS,Z         ; Skip if this digit is zero
0091 0499         00284          BCF      EGGNOW           ; Timer is not zero so clear egg alarm bit
0092 090D         00285          CALL     task_scan        ; Scan the next LED digit.
0093 0217         00286          MOVF     TMR_MIN_HD,W     ; Set the Z bit to check for zero
0094 0743         00287          BTFSS    STATUS,Z         ; Skip if this digit is zero
0095 0499         00288          BCF      EGGNOW           ; Timer is not zero so clear egg alarm bit
0096 0799         00289          BTFSS    EGGNOW           ; Skip if we are still at EGG Time
0097 05C3         00290          BSF      EGGOK            ; If we are not at EGG time, re-set egg alarm
0098 0779         00291          BTFSS    ALARMNOW         ; Skip if we are still at Alarm time
0099 05A3         00292          BSF      ALARMOK          ; If we are not at Alarm time, re-set alarm
009A 090D         00293          CALL     task_scan        ; Scan the next LED digit.
009B 0800         00294          RETLW    0
                  00295
                  00296 ; ****************************************
                  00297 ; * Increment the clock, timer or alarm  *
                  00298 ; ****************************************
009C              00299 inc_time
009C 0024         00300          MOVWF    FSR              ; Add one to clock second counter
009D 090D         00301          CALL     task_scan        ; Scan the next LED digit.
009E 02A0         00302          INCF     INDADDR,f        ; Add one to minute lower digit
009F 0C3C         00303          MOVLW    SEC_MAX          ; Place second max value into w
00A0 0080         00304          SUBWF    INDADDR,W        ; CLOCK_SEC - SEC_MAX -> W
00A1 0703         00305          BTFSS    STATUS,C         ; Skip if there is an overflow
00A2 0800         00306          RETLW    0                ; We are done so let's get out of here!
00A3 006B         00307          CLRF     CLK_SEC          ; Clear CLK_second counter
00A4 02A4         00308          INCF     FSR,F            ; Move to the next digit
00A5 02A0         00309          INCF     INDADDR,F        ; Add 1 to minute LOW digit
00A6 0AA9         00310          GOTO     skip_min_fsr     ; Jump to the next digit
00A7              00311 inc_min_ld
00A7 0024         00312          MOVWF    FSR
00A8 02A0         00313          INCF     INDADDR,F        ; Add 1 to minute LOW digit
00A9              00314 skip_min_fsr
00A9 090D         00315          CALL     task_scan        ; Scan the next LED digit.
00AA 0C0A         00316          MOVLW    MIN_LD_MAX       ; Place minute lower digit max value into W
```

AN615

```
00AB 0080    00317            SUBWF    INDADDR,W        ; CLK_MIN_LD - MIN_LD_MAX -> W
00AC 0703    00318            BTFSS    STATUS,C         ; Skip if there is an overflow
00AD 0800    00319            RETLW    0                ; We are done so let's get out of here!
00AE 0060    00320            CLRF     INDADDR          ; Clear CLK minute low digit
00AF 02A4    00321            INCF     FSR,F            ; Move to the minute high digit
00B0 02A0    00322            INCF     INDADDR,F        ; Add one to minute high digit
00B1         00323  inc_min_hd
00B1 090D    00324            CALL     task_scan        ; Scan the next LED digit.
00B2 0C06    00325            MOVLW    MIN_HD_MAX       ; Place minute high digit max value into W
00B3 0080    00326            SUBWF    INDADDR,W        ; CLK_MIN_HD - MIN_HD_MAX -> W
00B4 0703    00327            BTFSS    STATUS,C         ; Skip if there is an overflow
00B5 0800    00328            RETLW    0                ; We are done so let's get out of here!
00B6 0060    00329            CLRF     INDADDR          ; Clear CLK minute high digit
00B7 02A4    00330            INCF     FSR,F            ; Move to the hour low digit
00B8 02A0    00331            INCF     INDADDR,F        ; Add one to hour low digit
00B9 0ABE    00332            GOTO     skip_hour_fsr    ; Jump to the next digit
00BA         00333  inc_hour_ld
00BA 0024    00334            MOVWF    FSR
00BB 02A4    00335            INCF     FSR,F
00BC 02A4    00336            INCF     FSR,F
00BD 02A0    00337            INCF     INDADDR,F        ; Add 1 to minute LOW digit
00BE         00338  skip_hour_fsr
00BE 090D    00339            CALL     task_scan        ; Scan the next LED digit.
00BF 0C0A    00340            MOVLW    MIN_LD_MAX       ; Place hour lower digit max value into W
00C0 0080    00341            SUBWF    INDADDR,W        ; CLK_HOUR_LD - HOUR_LD_MAX -> W
00C1 0703    00342            BTFSS    STATUS,C         ; Skip if there is an overflow
00C2 0AC7    00343            GOTO     check_inc        ; We need to check for overflow
00C3 0060    00344            CLRF     INDADDR          ; Clear CLK hour low digit
00C4 02A4    00345            INCF     FSR,F            ; Move to the hour high digit
00C5 02A0    00346            INCF     INDADDR,F        ; Add one to hour high digit
00C6 0AC8    00347            GOTO     inc_hour_hd
00C7         00348  check_inc
00C7 02A4    00349            INCF     FSR,F            ; Move to hour high digit
00C8         00350  inc_hour_hd
00C8 090D    00351            CALL     task_scan        ; Scan the next LED digit.
00C9 0C02    00352            MOVLW    HOUR_HD_MAX      ; Place hour high digit max value into W
00CA 0639    00353            BTFSC    FLAGS,1
00CB 0ACE    00354            GOTO     off_mode1
00CC 0619    00355            BTFSC    FLAGS,0
00CD 0C09    00356            MOVLW    MIN_LD_MAX-1
00CE         00357  off_mode1
00CE 0080    00358            SUBWF    INDADDR,W        ; CLK_HOUR_HD - HOUR_HD_MAX -> W
00CF 0703    00359            BTFSS    STATUS,C         ; Skip if there is an overflow
00D0 0800    00360            RETLW    0                ; We are done so let's get out of here!
00D1 00E4    00361            DECF     FSR,F            ; Move to the hour low digit
00D2 090D    00362            CALL     task_scan        ; Scan the next LED digit.
00D3 0C04    00363            MOVLW    HOUR_LD_MAX      ; Place hour high digit max value into W
00D4 0639    00364            BTFSC    FLAGS,1
00D5 0AD8    00365            GOTO     off_mode2
00D6 0619    00366            BTFSC    FLAGS,0
00D7 0C00    00367            MOVLW    0                ; Clear W
00D8         00368  off_mode2
00D8 0080    00369            SUBWF    INDADDR,W        ; CLK_HOUR_HD - HOUR_HD_MAX -> W
00D9 0703    00370            BTFSS    STATUS,C         ; Skip if there is an overflow
00DA 0800    00371            RETLW    0                ; We are done so let's get out of here!
00DB 090D    00372            CALL     task_scan        ; Scan the next LED digit.
00DC 0060    00373            CLRF     INDADDR          ; Clear hour high digit
00DD 0639    00374            BTFSC    FLAGS,1
00DE 0AE0    00375            GOTO     off_mode3
00DF 0719    00376            BTFSS    FLAGS,0
00E0         00377  off_mode3
00E0 0000    00378            NOP
00E1 02A4    00379            INCF     FSR,F            ; Move to the hour high digit
00E2 0060    00380            CLRF     INDADDR          ; Clear one hour low digit
00E3 090D    00381            CALL     task_scan
00E4 0800    00382            RETLW    0                ; We are done so let's get out of here!
```

AN615

```
              00383
00E5          00384 dec_hour_ld
00E5 0AF9     00385        GOTO     dec_hour_ld_vect  ; ran out of CALL space....
              00386
              00387 ; ***************************************
              00388 ; * Decrement the clock, alarm or timer *
              00389 ; ***************************************
00E6          00390 dec_time
00E6          00391 dec_min_ld
00E6 0024     00392        MOVWF    FSR              ; Set up pointer for indirect address
00E7 090D     00393        CALL     task_scan        ; Scan the next LED digit.
00E8 00E0     00394        DECF     INDADDR,F        ; Subtract one from CLK_MIN_LD
00E9 0240     00395        COMF     INDADDR,W        ; Set the Z bit to check for zero
00EA 0743     00396        BTFSS    STATUS,Z         ; Skip if CLK_MIN_LD is zero
00EB 0800     00397        RETLW    0                ; We are done... Let's get out of here
00EC 0C09     00398        MOVLW    MIN_LD_MAX - 1   ; Place minute lower digit max value into W
00ED 0020     00399        MOVWF    INDADDR          ; MIN_LD_MAX -> CLK_MIN_LD
00EE          00400 dec_min_hd
00EE 090D     00401        CALL     task_scan        ; Scan the next LED digit.
00EF 02A4     00402        INCF     FSR,F            ; Move the pointer to Min HIGH DIGIT
00F0 00E0     00403        DECF     INDADDR,F        ; Subtract one from CLK_MIN_HD
00F1 0240     00404        COMF     INDADDR,W        ; Set the Z bit to check for zero
00F2 0743     00405        BTFSS    STATUS,Z         ; Skip if CLK_MIN_LD is zero
00F3 0800     00406        RETLW    0                ; We are done... Let's get out of here
00F4 0C05     00407        MOVLW    MIN_HD_MAX - 1   ; Place minute lower digit max value into W
00F5 0020     00408        MOVWF    INDADDR          ; MIN_HD_MAX -> CLK_MIN_HD
00F6 090D     00409        CALL     task_scan        ; Scan the next LED digit.
00F7 02A4     00410        INCF     FSR,F            ; Move the pointer to Hour LOW DIGIT
00F8 0AFD     00411        GOTO     skip_dhour_fsr   ; Jump to the next digit
00F9          00412 dec_hour_ld_vect
00F9 0024     00413        MOVWF    FSR
00FA 02A4     00414        INCF     FSR,F
00FB 02A4     00415        INCF     FSR,F
00FC 090D     00416        CALL     task_scan        ; Scan the next LED digit.
00FD          00417 skip_dhour_fsr
00FD 00E0     00418        DECF     INDADDR,F        ; Subtract one from CLK_HOUR_LD
00FE 0240     00419        COMF     INDADDR,W        ; Set the Z bit to check for zero
00FF 0743     00420        BTFSS    STATUS,Z         ; Skip if CLK_MIN_LD is zero
0100 0B06     00421        GOTO     check_hour
0101 0C09     00422        MOVLW    MIN_LD_MAX - 1   ; Place minute lower digit max value into W
0102 0020     00423        MOVWF    INDADDR          ; MIN_LD_MAX -> CLK_HOUR_LD
0103 02A4     00424        INCF     FSR,F            ; Move the pointer to Hour HIGH DIGIT
0104 00E0     00425        DECF     INDADDR,F        ; Subtract one from CLK_HOUR_HD
0105 0B07     00426        GOTO     dec_hour_hd
0106          00427 check_hour
0106 02A4     00428        INCF     FSR,F            ; Point to hour high digit
0107          00429 dec_hour_hd
0107 090D     00430        CALL     task_scan        ; Scan the next LED digit.
0108 0240     00431        COMF     INDADDR,W
0109 0743     00432        BTFSS    STATUS,Z
010A 0800     00433        RETLW    0
010B 090D     00434        CALL     task_scan        ; Scan the next LED digit.
010C 00E4     00435        DECF     FSR,F
010D 0C09     00436        MOVLW    .9               ; Reset digit to 9
010E 0080     00437        SUBWF    INDADDR,W
010F 0743     00438        BTFSS    STATUS,Z         ; Skip if CLK_MIN_LD is zero
0110 0800     00439        RETLW    0                ; We are done... Let's get out of here
0111 090D     00440        CALL     task_scan        ; Scan the next LED digit.
0112 02A4     00441        INCF     FSR,F
0113 0C02     00442        MOVLW    HOUR_HD_MAX      ; Place minute lower digit max value into W
0114 0739     00443        BTFSS    FLAGS,1          ; Skip if CLOCK or ALARM mode
0115 0C09     00444        MOVLW    .9               ; Reset digit to 9
0116 0020     00445        MOVWF    INDADDR          ; HOUR_HD_MAX -> CLK_HOUR_HD
0117 0C03     00446        MOVLW    HOUR_LD_MAX - 1  ; Place minute lower digit max value into W
0118 0739     00447        BTFSS    FLAGS,1          ; Skip if CLOCK or ALARM mode
0119 0C09     00448        MOVLW    .9               ; Reset digit to 9
```

AN615

```
011A 00E4    00449          DECF     FSR,F            ; Move the pointer to Min LOW DIGIT
011B 0020    00450          MOVWF    INDADDR          ; HOUR_LD_MAX -> CLK_HOUR_LD
011C 090D    00451          CALL     task_scan        ; Scan the next LED digit.
011D 0800    00452          RETLW    0                ; We are done... Let's get out of here
             00453
             00454 ; ****************************************
             00455 ; * Main loop calls all tasks as needed  *
             00456 ; ****************************************
011E         00457 main_loop
011E 090D    00458          CALL     task_scan        ; Scan the next LED digit.
011F 0201    00459          MOVF     TMR0,W           ; Place current TMR0 value into W
0120 019A    00460          XORWF    PREVTMR0,W       ; Lets see which bits have changed...
0121 003C    00461          MOVWF    TEMP             ; All changed bits are placed in temp for test
0122 01BA    00462          XORWF    PREVTMR0,F       ; Update Previous TMR0 value.
0123 07FC    00463          BTFSS    SECBIT           ; Skip if it is not time to increment second
0124 0B1E    00464          GOTO     main_loop        ; Go back to main loop if 250 mS not passed
0125 0C20    00465          MOVLW    b'00100000'      ; Bits 6 and 5 of FLAGS used as divide by 4
0126 01F9    00466          ADDWF    FLAGS,F          ; Add one to bit 5
0127 07F9    00467          BTFSS    TIMENOW          ; Check bit 7 - if four adds occur, skip
0128 0B38    00468          GOTO     skip_timer       ; One second has not passed - skip timers
0129 090D    00469          CALL     task_scan        ; Scan the next LED digit.
012A 04F9    00470          BCF      TIMENOW          ; Clear out second passed flag
012B 0C0B    00471          MOVLW    CLK_SEC          ; Place pointer to increment clock
012C 099C    00472          CALL     inc_time         ; Increment the clock
012D 0974    00473          CALL     check_time       ; Check for alarm or timer conditions
012E 0699    00474          BTFSC    EGGNOW           ; Do NOT decrease timer if zero
012F 0B38    00475          GOTO     skip_timer       ; Jump out if egg timer is zero
0130 06D8    00476          BTFSC    UPKEY            ; Skip if UP key is NOT pressed
0131 0B38    00477          GOTO     skip_timer       ; Jump out if UP key is pressed
0132 06B8    00478          BTFSC    DOWNKEY          ; Skip if DOWN key is NOT pressed
0133 0B38    00479          GOTO     skip_timer       ; Jump out if DOWN key is pressed
0134 0C14    00480          MOVLW    TMR_SEC_LD       ; Place pointer to decrement timer
0135 09E6    00481          CALL     dec_time         ; Decrement countdown timer
0136 0C28    00482          MOVLW    ALARMCYCCNT      ; Place the number of alarm beeps into W
0137 003F    00483          MOVWF    ALARMCNT         ; Move beep count to ALARMCNT
0138         00484 skip_timer
0138 07A3    00485          BTFSS    ALARMOK          ; Skip if this is the first pass into alarm
0139 0B3F    00486          GOTO     skip_wakeup      ; Second pass - do not re-init ALARMCNT
013A 0779    00487          BTFSS    ALARMNOW         ; Skip if this is alarm pass
013B 0B3F    00488          GOTO     skip_wakeup      ; Countdown timer - do not re-init ALARMCNT
013C 0C28    00489          MOVLW    ALARMCYCCNT      ; Place the number of alarm beeps into W
013D 003F    00490          MOVWF    ALARMCNT         ; Move beep count to ALARMCNT
013E 04A3    00491          BCF      ALARMOK          ; Clear flag for second pass
013F         00492 skip_wakeup
013F 090D    00493          CALL     task_scan        ; Scan the next LED digit.
0140 0679    00494          BTFSC    ALARMNOW         ; Skip if alarm clock is not set
0141 0B45    00495          GOTO     send_alarm       ; Blast out a beep
0142 0699    00496          BTFSC    EGGNOW           ; Skip if countdown timer is not alarming
0143 0B45    00497          GOTO     send_alarm       ; Blast out a beep
0144 0B4A    00498          GOTO     skip_alarm       ; Skip beeping and continue
0145         00499 send_alarm
0145 021F    00500          MOVF     ALARMCNT,W       ; Place ALARMCNT into W
0146 0643    00501          BTFSC    STATUS,Z         ; Skip if not zero
0147 0B4A    00502          GOTO     skip_alarm       ; We are done beeping - skip and continue
0148 02FF    00503          DECFSZ   ALARMCNT,F       ; Decrement beep count and skip when zero
0149 0906    00504          CALL     buzz_now         ; Blast out the beep!!!
014A         00505 skip_alarm
014A 07B9    00506          BTFSS    FLAGS,5          ; Skip if it is time to scan the keys 1/2 sec
014B 0B9A    00507          goto     finish_update    ; Jump to finish updates - don't scan
014C 0966    00508          CALL     scan_keys        ; Scan the keys and load value into KEYPAT
014D 090D    00509          CALL     task_scan        ; Scan the next LED digit.
014E 0798    00510          BTFSS    MODEKEY          ; Skip if the MODEKEY is pressed
014F 0B55    00511          GOTO     same_mode        ; Not pressed so it is the same mode...
0150 079C    00512          BTFSS    MODEKEYCHG       ; Skip if the is pressing edge
0151 0B55    00513          GOTO     same_mode        ; Button is held so it is the same mode...
0152 02B9    00514          INCF     FLAGS,F          ; Advance the mode by incrimenting bits 0,1
```

AN615

```
0153  0459    00515         BCF     FLAGS,2         ; Force mode to wrap-around by clearing bit 2
0154  0953    00516         CALL    turnon_scan     ; Mode button pressed - must turn on LEDs
              00517
0155          00518 same_mode
0155  090D    00519         call    task_scan       ; Scan the next LED digit.
0156  06D8    00520         BTFSC   UPKEY           ; Skip if the UP key is not pressed
0157  0B66    00521         GOTO    serve_up_key    ; UP key is pressed - jump to serve it!
0158  06B8    00522         BTFSC   DOWNKEY         ; Skip if the DOWN key is not pressed
0159  0B81    00523         GOTO    serve_down_key  ; DOWN key is pressed - jump to serve it!
015A  0C08    00524         MOVLW   INIT_MODE_COUNT ; UP and DOWN not pressed - re-init mode count
015B  003E    00525         MOVWF   MODE_COUNT      ; Change back to lower digits for setting
015C  023D    00526         MOVF    DISPONCNT,F     ; Update Z bit in STATUS reg display on time
015D  0743    00527         BTFSS   STATUS,Z        ; Skip if displays should be OFF
015E  00FD    00528         DECF    DISPONCNT,F     ; Decrement display ON counter
015F  0743    00529         BTFSS   STATUS,Z        ; Skip if displays should be OFF
0160  0B9A    00530         GOTO    finish_update   ; Displays are ON - jump to finish updates
0161  0419    00531         BCF     FLAGS,0         ; Restore the mode to displays OFF
0162  0439    00532         BCF     FLAGS,1         ; Restore the mode to displays OFF
0163  0066    00533         CLRF    PORTB           ; Clear out segment drives on PORTB
0164  0065    00534         CLRF    PORTA           ; Clear out common digit drives on PORTA
0165  0B9A    00535         GOTO    finish_update   ; Jump to finish updates
0166          00536 serve_up_key
0166  090D    00537         call    task_scan       ; Scan the next LED digit.
0167  0619    00538         BTFSC   FLAGS,0         ; Skip if not in TIMER or CLOCK mode
0168  0B6D    00539         GOTO    no_up_display   ; Currently in TIMER or CLOCK - keep mode
0169  0639    00540         BTFSC   FLAGS,1         ; Skip if not in ALARM mode
016A  0B6D    00541         GOTO    no_up_display   ; Currently in ALARM - keep mode
016B  0519    00542         BSF     FLAGS,0         ; Set to CLOCK mode
016C  0539    00543         BSF     FLAGS,1         ; Set to CLOCK mode
016D          00544 no_up_display
016D  007F    00545         CLRF    ALARMCNT        ; A key was pressed, so turn off alarm
016E  0953    00546         call    turnon_scan     ; Turn on the LEDs
016F  0798    00547         BTFSS   MODEKEY         ; Skip if MODE is pressed as well
0170  0B9A    00548         GOTO    finish_update   ; MODE is not pressed - jump to finish update
0171  021E    00549         MOVF    MODE_COUNT,W    ; Update STATUS Z bit for mode count
0172  0743    00550         BTFSS   STATUS,Z        ; Skip if we have counted down to zero
0173  00FE    00551         DECF    MODE_COUNT,F    ; Decrement the mode count
0174  090D    00552         call    task_scan       ; Scan the next LED digit.
0175  021E    00553         MOVF    MODE_COUNT,W    ; Update the Z bit to check for zero
0176  0743    00554         BTFSS   STATUS,Z        ; Skip if we have incremented for 7 times
0177  0B7C    00555         GOTO    serve_min_up    ; Incriment the minutes digits
0178  00D9    00556         DECF    FLAGS,W         ; Place current mode into W
0179  0900    00557         CALL    mode_timer      ; Look-up register RAM address for current mode
017A  0BBA    00558         CALL    inc_hour_ld     ; Add one hour to the current display
017B  0B9A    00559         GOTO    finish_update   ; Jump to finish updates
017C          00560 serve_min_up
017C  090D    00561         call    task_scan       ; Scan the next LED digit.
017D  00D9    00562         DECF    FLAGS,W         ; Place current mode into W
017E  0900    00563         CALL    mode_timer      ; Look-up register RAM address for current mode
017F  09A7    00564         CALL    inc_min_ld      ; Add one minute to the current display
0180  0B9A    00565         GOTO    finish_update   ; Jump to finish updates
0181          00566 serve_down_key
0181  090D    00567         call    task_scan       ; Scan the next LED digit.
0182  0619    00568         BTFSC   FLAGS,0         ; Skip if not in TIMER or CLOCK mode
0183  0B88    00569         GOTO    no_dn_display   ; Currently in TIMER or CLOCK - keep mode
0184  0639    00570         BTFSC   FLAGS,1         ; Skip if not in ALARM mode
0185  0B88    00571         GOTO    no_dn_display   ; Currently in ALARM - keep mode
0186  0519    00572         BSF     FLAGS,0         ; Set to CLOCK mode
0187  0539    00573         BSF     FLAGS,1         ; Set to CLOCK mode
0188          00574 no_dn_display
0188  007F    00575         CLRF    ALARMCNT        ; A key was pressed, so turn off alarm
0189  0953    00576         CALL    turnon_scan     ; Turn on the LEDs
018A  0798    00577         BTFSS   MODEKEY         ; Skip if MODE is pressed as well
018B  0B9A    00578         GOTO    finish_update   ; MODE is not pressed - jump to finish update
018C  021E    00579         MOVF    MODE_COUNT,W    ; Update STATUS Z bit for mode count
018D  0743    00580         BTFSS   STATUS,Z        ; Skip if we have counted down to zero
```

AN615

```
018E 00FE   00581          DECF    MODE_COUNT,F    ; Decrement the mode count
            00582
018F 090D   00583          call    task_scan       ; Scan the next LED digit.
0190 021E   00584          MOVF    MODE_COUNT,W    ; Update the Z bit to check for zero
0191 0743   00585          BTFSS   STATUS,Z        ; Skip if we have incrimented for 7 times
0192 0B97   00586          GOTO    serve_min_down  ; Decrement the minutes digits
0193 00D9   00587          DECF    FLAGS,W         ; Place current mode into W
0194 0900   00588          CALL    mode_timer      ; Look-up register RAM address for current mode
0195 09E5   00589          CALL    dec_hour_ld     ; Subtract one hour from the current display
0196 0B9A   00590          GOTO    finish_update   ; Jump to finish updates
0197        00591 serve_min_down
0197 00D9   00592          DECF    FLAGS,W         ; Place current mode into W
0198 0900   00593          CALL    mode_timer      ; Look-up register RAM address for current mode
0199 09E6   00594          CALL    dec_min_ld      ; Subtract one minute from the current display
019A        00595 finish_update
019A 090D   00596          call    task_scan       ; Scan the next LED digit.
019B 0619   00597          BTFSC   FLAGS,0         ; Skip if in mode OFF or ALARM
019C 0BA4   00598          GOTO    new_display     ; Jump to update LED display registers
019D 0639   00599          BTFSC   FLAGS,1         ; Skip if in mode OFF
019E 0BA4   00600          GOTO    new_display     ; Jump to update LED display registers
019F 0067   00601          CLRF    DISPSEGS_A      ; Clear display regs to Shut off LED display
01A0 0068   00602          CLRF    DISPSEGS_B      ; Clear display regs to Shut off LED display
01A1 0069   00603          CLRF    DISPSEGS_C      ; Clear display regs to Shut off LED display
01A2 006A   00604          CLRF    DISPSEGS_D      ; Clear display regs to Shut off LED display
01A3 0B1E   00605          GOTO    main_loop       ; We are done - go back and do it again!
01A4        00606 new_display
01A4 00D9   00607          DECF    FLAGS,W         ; Move current mode state into W
01A5 0900   00608          CALL    mode_timer      ; Look-up register address of value to display
01A6 091F   00609          CALL    disp_value      ; Update display registers with new values
01A7 0B1E   00610          GOTO    main_loop       ; We are done - go back and do it again!
            00611
            00612 ; ****************************************
            00613 ; * Set up and initialize the processor  *
            00614 ; ****************************************
01A8        00615 init
01A8 0C03   00616          MOVLW   OPTION_SETUP    ; Place option reg setup into W
01A9 0002   00617          OPTION                  ; Set up OPTION register
01AA 0C05   00618          MOVLW   PORTA           ; Place beginning of RAM/Port location into W
01AB 0024   00619          MOVWF   FSR             ; Now initialize FSR with this location
01AC        00620 clear_mem
01AC 0060   00621          CLRF    INDADDR         ; Clear the FSR pointed memory location
01AD 03E4   00622          INCFSZ  FSR,F           ; Point to the next location
01AE 0BAC   00623          GOTO    clear_mem       ; Jump back to clear memory routine
01AF 0572   00624          BSF     ALM_HOUR_LD,3   ; Place 8:00 into alarm register
01B0 02AE   00625          INCF    CLK_HOUR_LD,F   ; Place 1:00 into clock register
01B1 0CEE   00626          MOVLW   0EEh            ; Turn on display A scan line, others off
01B2 003B   00627          MOVWF   PREVSCAN        ;
01B3 0040   00628          CLRW                    ;
01B4 0006   00629          TRIS    PORTB           ; Make all Port B pins outputs.
01B5 0005   00630          TRIS    PORTA           ; Make all Port A pins outputs.
01B6 0539   00631          BSF     FLAGS,1         ; Set up current mode to CLOCK, display ON
01B7 0519   00632          BSF     FLAGS,0
01B8 04A3   00633          BCF     ALARMOK         ; Don't want to trigger alarms
01B9 04C3   00634          BCF     EGGOK
01BA 059D   00635          BSF     DISPON          ; Turn on the displays
01BB        00636 mfg_checkkey
01BB 0966   00637          CALL    scan_keys       ; Lets see what is pressed
01BC 07D8   00638          BTFSS   UPKEY           ; Goto self-test if UP key is pressed at pwr up
01BD 0B1E   00639          GOTO    main_loop       ; Normal operation - Jump to the main loop
            00640
            00641 ; ***************************************************************
            00642 ; * Self-test code for manufacturing only - test buttons and LEDs *
            00643 ; ***************************************************************
01BE        00644 mfg_selftest
01BE 0C70   00645          MOVLW   b'01110000'     ; Place all key on pattern into W
01BF 002D   00646          MOVWF   CLK_MIN_HD      ; Use CLK_MIN_HD for keystuck ON test
```

AN615

```
01C0 006F    00647          CLRF    CLK_HOUR_HD    ; Use CLK_HOUR_HD for keystuck OFF test
01C1         00648 mfg_display
01C1 020B    00649          MOVF    CLK_SEC,W      ; Current segment display count -> W
01C2 0949    00650          CALL    mfg_led_lookup ; Look-up the next segment pattern to display
01C3 0026    00651          MOVWF   PORTB          ; Move the pattern to PORT B to display it
01C4         00652 mfg_timer
01C4 0201    00653          MOVF    TMR0,W         ; Place current TMR0 value into W
01C5 019A    00654          XORWF   PREVTMR0,W     ; Lets see which bits have changed...
01C6 003C    00655          MOVWF   TEMP           ; All changed bits are placed in temp for test
01C7 01BA    00656          XORWF   PREVTMR0,F     ; Update Previous TMR0 value.
01C8 07FC    00657          BTFSS   TEMP,7         ; Skip if it is not time to increment second
01C9 0BC4    00658          GOTO    mfg_timer      ; It is not time to move to next digit - go back
01CA 02AB    00659          INCF    CLK_SEC,F      ; Move to the next display pattern
01CB         00660 mfg_check_digit
01CB 07AB    00661          BTFSS   CLK_SEC,5      ; Skip if we have timed out waiting for button
01CC 0BD5    00662          GOTO    mfg_doneclk    ; Jump to check for the next button press
01CD         00663 mfg_nextdigit
01CD 006B    00664          CLRF    CLK_SEC        ; Clear out timer
01CE 0906    00665          CALL    buzz_now       ; Send out a buzzer beep!
01CF 077B    00666          BTFSS   PREVSCAN,3     ; Skip if we have NOT tested the last digit
01D0 0BE5    00667          GOTO    finish_mfg_test ; Jump to the end after last digit tested
01D1 035B    00668          RLF     PREVSCAN,W     ; Select the next digit through a rotate..
01D2 037B    00669          RLF     PREVSCAN,F
01D3 021B    00670          MOVF    PREVSCAN,W     ; Place next digit select into W
01D4 0025    00671          MOVWF   PORTA          ; Update port A to select next digit
01D5         00672 mfg_doneclk
01D5 0966    00673          CALL    scan_keys      ; Scan the keys to see what is pressed...
01D6 0218    00674          MOVF    KEYPAT,W       ; Place pattern into W
01D7 016D    00675          ANDWF   CLK_MIN_HD,F   ; Make sure keys are not stuck ON
01D8 012F    00676          IORWF   CLK_HOUR_HD,F  ; Make sure each key is pressed at least once
01D9 077B    00677          BTFSS   PREVSCAN,3     ; Skip if we are NOT at the last digit
01DA 05F8    00678          BSF     KEYPAT,7       ; Set flag bit to indicate we are done!
01DB 0C08    00679          MOVLW   .8             ; Place 8 into W
01DC 008B    00680          SUBWF   CLK_SEC,W      ; CLK_SEC - W => W
01DD 0703    00681          BTFSS   STATUS,C
01DE 0078    00682          CLRF    KEYPAT
01DF 03B8    00683          SWAPF   KEYPAT,F
01E0 025B    00684          COMF    PREVSCAN,W
01E1 0158    00685          ANDWF   KEYPAT,W
01E2 0743    00686          BTFSS   STATUS,Z
01E3 0BCD    00687          GOTO    mfg_nextdigit
01E4 0BC1    00688          GOTO    mfg_display
01E5         00689 finish_mfg_test
01E5 022D    00690          MOVF    CLK_MIN_HD,F
01E6 0743    00691          BTFSS   STATUS,Z
01E7 0BEF    00692          GOTO    bad_switch
01E8 020F    00693          MOVF    CLK_HOUR_HD,W
01E9 0F70    00694          XORLW   070h
01EA 0743    00695          BTFSS   STATUS,Z
01EB 0BEF    00696          GOTO    bad_switch
01EC         00697 mfg_cleanup
01EC 006F    00698          CLRF    CLK_HOUR_HD    ; Restore temp registers to zero
01ED 006D    00699          CLRF    CLK_MIN_HD     ; Restore temp registers to zero
01EE 0B1E    00700          GOTO    main_loop      ; Jump to main loop
01EF         00701 bad_switch
01EF 026D    00702          COMF    CLK_MIN_HD,F
01F0 038D    00703          SWAPF   CLK_MIN_HD,W
01F1 0038    00704          MOVWF   KEYPAT
01F2 05EF    00705          BSF     CLK_HOUR_HD,7
01F3 038F    00706          SWAPF   CLK_HOUR_HD,W
01F4 0178    00707          ANDWF   KEYPAT,F
01F5 0C7F    00708          MOVLW   07Fh
01F6 0026    00709          MOVWF   PORTB
01F7 006C    00710          CLRF    CLK_MIN_LD
01F8 05AC    00711          BSF     CLK_MIN_LD,5
01F9         00712 loop_bad_sw
```

AN615

```
01F9 0907    00713    CALL    buzz_now_dispon ; Beep the buzzer constantly for a few secs
01FA 02EC    00714    DECFSZ  CLK_MIN_LD,F    ; Decrement counter and skip when done
01FB 0BF9    00715    GOTO    loop_bad_sw     ; Not done buzzing - go back and do it again
01FC 0BEC    00716    GOTO    mfg_cleanup     ; Done buzzing - clean-up and run clock
             00717    END
```

```
MEMORY USAGE MAP ('X' = Used,  '-' = Unused)

0000 : XXXXXXXXXXXXXXXX XXXXXXXXXXXXXXXX XXXXXXXXXXXXXXXX XXXXXXXXXXXXXXXX
0040 : XXXXXXXXXXXXXXXX XXXXXXXXXXXXXXXX XXXXXXXXXXXXXXXX XXXXXXXXXXXXXXXX
0080 : XXXXXXXXXXXXXXXX XXXXXXXXXXXXXXXX XXXXXXXXXXXXXXXX XXXXXXXXXXXXXXXX
00C0 : XXXXXXXXXXXXXXXX XXXXXXXXXXXXXXXX XXXXXXXXXXXXXXXX XXXXXXXXXXXXXXXX
0100 : XXXXXXXXXXXXXXXX XXXXXXXXXXXXXXXX XXXXXXXXXXXXXXXX XXXXXXXXXXXXXXXX
0140 : XXXXXXXXXXXXXXXX XXXXXXXXXXXXXXXX XXXXXXXXXXXXXXXX XXXXXXXXXXXXXXXX
0180 : XXXXXXXXXXXXXXXX XXXXXXXXXXXXXXXX XXXXXXXXXXXXXXXX XXXXXXXXXXXXXXXX
01C0 : XXXXXXXXXXXXXXXX XXXXXXXXXXXXXXXX XXXXXXXXXXXXXXXX XXXXXXXXXXXXX--X
0F80 : ---------------- ---------------- ---------------- ----------------
0FC0 : ---------------- ---------------- ---------------- --------------X

All other memory blocks unused.


Errors   :    0
Warnings :    0
Messages :    0
```

CHAPTER 10

ADDING EXTERNAL STATIC RAM TO THE PIC16C74

INTRODUCTION

Here is something that you will like! Using this procedure from Microchip's application note TB011 (*Figure
10.5*), you will be able to add external RAM to any project
involving the PIC16C74 microcontroller. While the internal RAM with most PICMICRO® MCUs is more than adequate for many endeavors, there will be times when
additional RAM is necessary. Examples of this might be
projects that require very large lookup tables or video
circuits that need abundant buffers.

However, these situations can be handily addressed by
following the methods described in this chapter. And the
great thing about this is simplicity! Nothing difficult or
menacing about it! So, next time you want to design a
project around a PICMICRO® microcontroller but don't
have enough internal memory for your program, think
about Chapter 10! This will likely be your ticket.

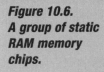

Figure 10.6.
A group of static
RAM memory
chips.

THEORY

As stated above, there is going to be that time when the internal random access memory (RAM) of the PICMICRO® microcontrollers just isn't enough for the task. Actually, I haven't run into many of these times, but trust me, they do exist. And this might not be a situation where the software itself is too big. Microchip points out digital voice storage as an excellent example.

But, for whatever reason, the project is going to need more RAM. Never fear, the answer to this dilemma is at hand! With some simple code and a few external chips (see *Figure 10.6*), additional memory can easily be achieved and at a remarkably inexpensive cost. Due to the speed of most applications, expensive FSRAM or a fast address/data bus are not necessary. This, of course, is good news to any designer, especially if the project may end up a product.

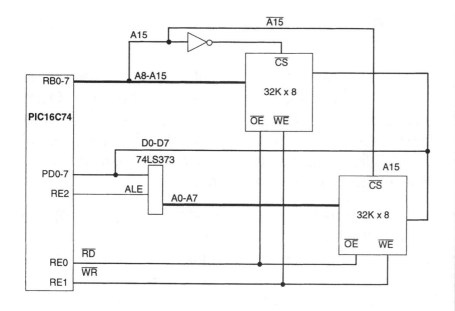

Figure 10.1.
Block diagram for
adding static
RAM memory to
PIC16CXXX
projects.

IMPLEMENTATION

Looking at *Figure 10.1*, a block diagram of this scheme, it can be seen that this is very similar to the method used by some of the older microprocessors for memory expansion. That is, multiplexing the address/data bus. The PIC16C74 does not have an external bus for address and data transfer, however, so Microchip cleverly uses the input/output (I/O) ports to make up the requisite 19 bus lines (16 for address/data and three for control). This is a departure from older systems, but serves this application well.

In this approach, the PIC16C74 runs at 20 megahertz, which does make the system faster than the 4- or 10-megahertz speeds normally encountered. Naturally, this must be reflected in the software. However, using software to achieve the address/data busing is more than acceptable for many projects.

Okay, let's take a closer look at this. Again referring to Figure 10.1, you will see that lines 0 to 7 of PORTD are used for the multiplexed lower address/data bus. These become address/data lines 0 through 7, while the eight PORTB lines are 8 to 15. This half of the bus is not multiplexed, but it doesn't need to be.

PORTE contributes the read (RD) line (PORTE 0), the write (WR) line (PORTE 1), and address latch enable (ALE) line (PORTE 2). The data is stored and/or retrieved from one of two 32K x 8 Static RAM chips (SRAM) and the app notes recommend the Fujitsu MB84256C-70 as prospective memory. More will be said about this in a moment.

The address range of one memory chip is 00007h to 7FFFh, while the second chip's range is 8000h to FFFFh. Individual chip selection is achieved using A15 and "not" A15. And, last, but not least, the 74LS373 is used to demultiplex the lower address/data bus. This is accomplished in the usual fashion using the address latch enable (ALE). See *Figure 10.2* for the bus timing.

Figure 10.2.
Read and write
cycle bus timing.

Timing	Description	Minimum	Maximum
TCY	Instruction cycle time @ 20 MHz	200 ns	DC
TLLLH	ALE pulse width	1 TCY	—
TAVDV	Address valid to data valid	—	7 TCY
TRLDV	Read low to data valid	1 TCY	—
TRHDZ	Read high to data float	0	1 TCY
TWLWH	WRITE pulse width	1 TCY	—
TWHDX	WRITE high to data no longer valid (data hold time)	—	2 TCY
TAVWH	Address valid to write high	—	5 TCY

With the basic hardware out of the way, let's take a gander at the software. Initializing the ports is the job of subroutine "init_muxbus", and a quick look at Figure 10.5, the code, reveals the initial states of the address line and bus control signals in the code comments. This is pretty self-explanatory.

READ CYCLE

With that said, let me move on to the read cycle. First though, allow me to make a comment. I think you will notice that all the software is quite fundamental and very easy to write and use. This is one of many factors that make the PICMICRO® MCU line of microcontrollers such a pleasure to employ. Hence, if you haven't already worked with these devices, you are in for a surprise.

Okay, back to the code! In the read cycle, the 16-bit addresses are placed on the bus, and then, when ALE goes low, the 74LS373 latches the lower lines (0 to 7) on the memory chip. Next, the lower bus lines are changed to inputs and the read line (RD) is set low. This, in turn, sets the memory chip output buffers to "on".

The memory data is now placed on data lines 0 to 7, RD goes high, and memory output buffers are disabled. The lower data lines are returned to the output state, and data is read from all 16 lines. The read cycle is emulated on PORTB and PORTD with the subroutine "read_extmem", as seen in Figure 10.5.

Figure 10.3 sets out some parameters involved in all this, and it also illustrates that slow SRAM is very appropriate

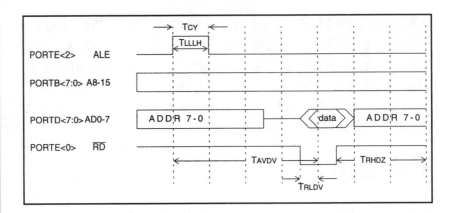

*Figure 10.3
Read cycle on
the multiplexed
address/data
bus.*

for this application. As Microchip points out, there are
three critical specifications regarding SRAM read cycles.
They are as follows.

1) Address Access Time (TACC)
2) Output Enable Time (TOE)
3) Data Float Time (TDF)

The address access time (TACC) coincides with the
"Address Valid to Data Valid" (TAVDV) timing of the
multiplexed (muxed) bus which, by the way, is 1.6
microseconds. Returning to the recommended Fujitsu
MB84256C-70 memory chips, they have a 70-
nanosecond access time and a 35-nanosecond output
enable time (TOE). This last value comes from the Fujitsu
data sheet. Hence, the emulated bus "Read Low to Data
Valid" timing, 200 nanoseconds, concurs with the TOE
of the memory ICs.

The last parameter, data float time (TDF), is only of
concern when back-to-back bus cycles are involved. So,
in this case, it is not necessary to consider this value. All

in all, this is—as much as I hate to use this tired phrase—fairly "user friendly" in its nature. But wait, the write cycle is even easier.

WRITE CYCLE

As you might expect, the write cycle places data into the external memory chips. This is done using the 16 address lines from the PIC16C74 and two of the control lines. Following the placement of the address, the ALE is brought first high, then low to latch the lower order bits. Next, the data is moved to the lower lines, the write (WR) is taken low then high and on the high pulse, the data is written into the memory. See what I mean? Eeeeeasy, and *Figure 10.4* illustrates the bus timing of this operation.

As for the code, it is also even simpler than the read cycle. Referring to Figure 10.5, notice the clarity of the "write_extmem" subroutine that handles this procedure. Thus, the entire process of reading and writing to an external memory device is a surprisingly pleasant task.

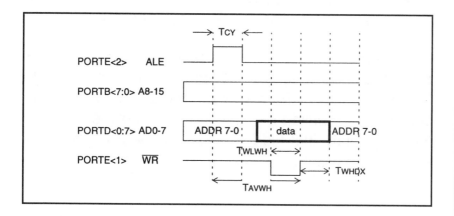

**Figure 10.4.
Write cycle bus
timing.**

And, following this same scheme, external memory could be added to other PICMICRO® microcontrollers.

On a final note regarding the write cycle, the "Write Pulse Width" (TWLWH) is an important parameter. For the PIC16C74 emulated address/data bus, as seen in Figure 10.2, this pulse width is one timing cycle (1 Tcy).

Now, this is a large value compared to the actual bus execution, but does serve to meet the specification for "Address Valid to Write High" (TAVWH) for most memory chips now available. And that is relatively slow compared to access timing with Static Random Access Memory (SRAM) or Electrically Programmable Read Only Memory (EPROM) devices.

CONCLUSION

So there you have a simple way to add external SRAM to any PIC16C74 project. This can be applied to any 40-pin PIC16CXXX device as well. As we have discussed, the procedure involves 16 address lines and eight data lines, with the lower lines being multiplexed to handle both address and data information. The result being the external data is placed in two 32K x 8 SRAMs.

Another nice factor of this approach is that various types of memory devices can be employed. In addition to SRAM, "Flash", EPROMs, or other parallel bus memory can easily be pulled into service. You know, sort of like the draft! Nah, just kidding! But, it is neat to have this kind of versatility!

Have fun with this one! It will be of immense help whenever your project design requires a large amount of data storage space. As you work with the Microchip ICs, you will learn how to do an amazing number of things you probably didn't think you could do with them. And often, some extra memory is just what the PC doctor ordered.

As you work with the Microchip ICs, you will learn how to do an amazing number of things you probably didn't think you could do with them.

TB011

APPENDIX A:PROGRAM LISTING: EXTERNAL MEMORY SUBROUTINES

```
;*********************************************************************
;*            Initialize the Multiplexed Address/Data Bus
;*
;* AD0-AD7 is PORTD 0-7
;* A8-A15 is PORTB 0-7
;* ALE is PORTE.2
;* RD# is PORTE.0
;* WR# is PORTE.1
;*
;* This init routine sets the multiplexed address/data bus up as
;* A0-A15 --> output low
;* ALE --> OUTPUT LOW
;* RD#,WR# --> OUTPUT HIGH
;*
;*********************************************************************

Init_MUXBUS
        bsf     STATUS,RP0          ;switch to bank 1 registers
        clrf    TRISB               ;set A8-A15 as output
        clrf    TRISD               ;set AD0-AD7 as output
        movlw   0xf8
        andwf   TRISE,F             ;ALE,RD#,WR# as output

        bcf     STATUS,RP0          ;switch to bank 0 registers

        clrf    ADHIGH              ;set A8-A15 to 0 (PORTB)
        clrf    ADLOW               ;set AD0-AD7 to 0 (PORTD)
        movlw   3                   ;ale=0,rd#=1,wr#=1
        movwf   PORTE
    return


;*********************************************************************
;*            Read External Memory of muxed bus
;*
;* INPUT: PORTB =A8-A15, PORTD = AD0-AD7
;* OUTPUT: W reg contains 8-bit data read from ext. mem.
;* CHANGED: W reg, ALE, RD#
;* (This READ routine has been modified to save the low order
;* address before a READ is done. The data read from memory will
;* destroy the address. After the read is done, the low order address
;* is written back out to PORTD.)
;*
;*********************************************************************
read_extmem
        movf    ADLOW,W             ;save low order address
        movwf   ADLOW_IMAGE

        bsf     PORTE,ALE           ;ALE high for 200ns, RD#, WR# low
        bcf     PORTE,ALE           ;ALE goes low (A0-7 latched)
        bsf     STATUS,RP0
        movlw   0xff
        movwf   TRISD               ;make PORTD input
        bcf     STATUS,RP0

        bcf     PORTE,RD            ;drop READ low

        movf    ADLOW,W             ;move read data from AD bus to w reg
        bsf     PORTE,RD            ;pull READ high (RD pulse is 400ns)
        bsf     STATUS,RP0
        clrf    TRISD               ;make PORTD (ADLOW) output again
        bcf     STATUS,RP0

        movwf   w_image             ;save READ data
        movf    ADLOW_IMAGE,W       ;restore low order address
```

Figure 10.5.
(Continued on next page) **Code for adding static RAM memory to PIC16CXXX projects.**

© 1997 Microchip Technology Inc.

TB011

```
        movwf   ADLOW           ;on port
        movf    w_image,W       ;restore READ data to w

        return
;*******************************************************************
;*              Write to External Memory on muxed bus
;* INPUT:   PORTB= A8-A15, PORTD = AD0-AD7, W= 8-bit data to write
;* OUTPUT: NOTHING
;* CHANGED: PORTE IS TOGGLED FOR ALE,WR# AND PUT BACK TO 011B
;* (This WRITE routine has been modified to save the low order
;*  address before a write is done. Then the low order address
;*  is put back on PORTD after the write.)
;*******************************************************************
write_extmem
        movwf   w_image         ;save w (data to write)
        movf    ADLOW,W
        movwf   ADLOW_IMAGE     ;save the low order address
        movf    w_image,W       ;restore w (data to write)

        bsf     PORTE,ALE       ;ALE high for 200ns, RD#,WR# low
        bcf     PORTE,ALE       ;latch lower address
        movwf   ADLOW           ;move write data to AD0-7
        bcf     PORTE,WR        ;WR# low for 200ns
        bsf     PORTE,WR        ;latch data in external memory

        movf    ADLOW_IMAGE,W
        movwf   ADLOW           ;restore low order address

        return
```

CHAPTER 11
THE PICMICRO®
MICROCONTROLLER
SOFTWARE

INTRODUCTION

As has been said numerous times in this book, with the majority of projects, the software does most of the work. Naturally, the ol' chicken-or-egg routine will rear its head whenever people discuss computers—that is, which came first, the software or the hardware.

In reality, however, both are equally important, as without the hardware, the software has nowhere to run. On the other hand, the hardware is nothing more than a pile of electronic components without the software.

But, the debate as to which is the most significant continues to rage on, with both sides having a well-defined agenda. That's not what this chapter is about, however, so allow me get back on track. At this point, we need to examine the software, and the tools to use it, in more detail, and that should not be taken as an affront to the hardware. They really do need each other, so we need to understand both of them equally.

The hardware is nothing more than a pile of electronic components without the software.

It is as simple a set of commands as I have ever seen regarding microcontrollers and/or microprocessors.

With that said, let's delve into the very pragmatic and efficient code that controls the PICMICRO® microcontrollers. We have already touched lightly on this subject, especially the instruction set, but to further your knowledge concerning these chips, a more in-depth analysis is in order.

INSTRUCTION SET

As I just said, the instruction set has been discussed, but I think this next section will clarify it even more. It is as simple a set of commands as I have ever seen regarding microcontrollers and/or microprocessors, and with the aid of the development tools, which I will cover, is as easy to use as any set I have ever encountered. And that all adds up to great news, at least as far as I'm concerned.

Additionally, the layout used to tell the assembler what to do with the text you create is quite straightforward in nature. Again, this makes working with the PICMICRO® MCUs a refreshing change from some of the earlier devices. I must admit that I have always been captivated by the microprocessor. The ability of electronic engineers to put into a small ceramic or plastic case (that can be held in your hand) what used to take an entire office to house is nothing short of genius to me. Maybe I'm being overenthusiastic, but I can't help it. These things are great!

Of course, I am supposed to be writing about software. Never fear, though, as that takes us right to the code. One of the reasons the present-day processors can be so compact is the software, and its ability do such a proficient job. And, with the PICMICRO® MCUs, the engineers have outdone themselves. This really is some of the best software around.

All right, let's go through the instruction set again command by command. Each one has a special purpose in the scheme of things and can be used to streamline the programs that run in the PICMICRO® MCUs. Hence, it is important to have a solid comprehension of the instruction set.

Since the various registers are so significant in making everything work the way you want it to, I will start with the 10 commands that affect, in one way or another, operation of the registers. Due to the comparatively simple nature of the PICMICRO® microcontrollers, moving data around inside the device is a necessity. And that is where the registers come into play. They also are important when it comes time to move data out of the device via the ports.

THE INSTRUCTION SET

CLRF, f: This is used to clear a selected register to the value of "0". That then allows information to be entered into the selected register.

CLRW: This one clears the "W" register ("working" register) to "0". As you work with PICMICRO® microcontroller programs, you will soon recognize the importance of the W register. For example, there is no direct way to store data in memory, so the data has to first be put into the W register, then moved to the memory. Of course, it is essential to first clear the W register.

BCF, f,b: Here, a chosen individual bit can be cleared to "0" in a selected register. This allows for greater manipulation of the data placed in the register.

BSF, f,b: This is another way to maneuver a register's data. In this command, a selected bit, within a certain register, can be changed to "1".

Here's an explanation of the 10 commands that affect, in one way or another, operation of the registers.

COMF, f,d: This instruction places the "compliment" of a register's contents into register W or register f. In other word, it changes all the 0's to 1's, and vice versa, then puts that information in the W or f ("file") register.

INCF, f,d: Using this instruction is an easy way to get from the last address "0xFF" to the first address "0x00". The selected register is incremented, or moved forward one, and the results end up in either register W or register f.

DECF, f,d: Here, the contents of a particular register is decremented, or decreased by one. This is an easy way to get from the first address "0x00" to the last address "0xFF. The end result goes in either register W or register f.

RRF, f,d: This command allows you to rotate a bit one place to the right, and the bit will rotate through the carry flag. Here again, the result will be placed in the W register or the f register (I think you can probably begin to see the significance of the W and f registers).

RLF, f,d: This does the same thing as the previous instruction, but rotates the bit, through the carry flag, one place to the left. Results are the same—the W or f register.

SWAPF, f,d: This one provides a way to actually exchange the most significant (MS) and least significant (LS) nibbles of data. Again, another refinement of data manipulation, and the outcome goes to either the W or f register.

Next, let's look at the three commands that handle the task of either defining or moving data.

Next, let's look at the three commands that handle the task of either defining or moving data. Naturally, these provide the means to get the data to where you want it and/or where it is needed.

MOVF, f.d: Here, a copy of the data in a particular register is moved into either register W or register f. Remember that data can't be moved directly into a memory address. It first has to be "moved" to the W register, then into the memory location.

MOVWF, f: This one moves a copy of the W register into a selected register. I think the significance of this command is self-explanatory.

MOVLW, f: This is a really handy command for delay or loop routines, as it loads the W register with a particular "literal" value. For example, in a delay routine, the literal value will determine the duration of the delay. Useful instruction!

The next set of instructions concerns controlling the microcontroller. There are but four of these, and they manage some important processor tasks.

CLRWDT: This command does several different things. For example, it will reset both the "watchdog timer" itself to zero (clear it), and reset the timer's prescaler. Additionally, it will set the "TO" and "PD" status bits. So, this is a useful instruction.

SLEEP: As you might expect from its name, this puts the microcontroller to "sleep". That is, it places it in a very low-power mode by keeping the essentials running while turning off some of the "whistles and bells" until they are needed. When you need to "wake up" the chip, this can be done with the reset, with an external input or with the watchdog timer.

TRIS, f: For a selected port, this command determines the port line status on a line-by-line basis. That is, whether the lines are inputs or outputs.

OPTION: Here, the contents of the W register are used to manage such things as the real-time clock trigger edge, the real-time clock

The next set of instructions concerns controlling the microcontroller.

Now, let's look at the instructions that control the flow of the program.

source, or the prescaler's ratio by placing the W register's data into the "option" register. You will find this one handy with projects that need to divide a value or "prescale" it.

Now, let's look at the instructions that control the flow of the program. These, of course, are used to keep the data moving smoothly within the microcontroller.

RETURN: This one is easy. It merely brings the program back to where it left off after handling a subroutine. Subroutines might be a loop or a delay routine, as well as others.

RETFIE: Here, the program is returned from an interrupt. Interrupts, as you might recall, are used to halt the microprocessor in midstream, that is after it has completed the task it is working on when the interrupt is called. Another task is then performed, and when that is completed, this command brings the program back.

RETLW, k: This is a handy instruction that loads the results of subroutine action into the W register. That happens when the program returns from said subroutine.

INCFSZ, f,d: Another increment command, but with a twist. Here, the particular register is incremented, but if the register content is "0", the next command is skipped. The result is placed in the W or f register. This one has all kinds of possibilities.

DECFSZ, f,d: Like the instruction above, this is another handy command. Instead of incrementing the register, it is decremented. And, if the register content is "0", the next instruction is skipped. Again, the result goes to register W or f.

GOTO, k: A simple but highly pragmatic command, "goto" takes the program to a specific address. You will see, and use, this instruction numerous times.

CALL, k: Another staple command, "call" brings up a subroutine at a specific starting point, or address. Like "goto", this one will become an old friend.

BTFSC, f,b: Here, a particular register bit is tested, and if the bit is "0", the next command is ignored. Like the "incfsz" and "decfsz" instructions, this and the one to follow are quite useful under certain conditions.

BTFSS, f,b: In this case, the particular register bit is tested, and if the bit is a "1", the next command is ignored. Another highly practical addition to the instruction set.

NOP: While this is not exactly a flow-control instruction, I am adding it to this section because it does involve data movement, or in this case, lack of it. NOP stands for "nothing", and that is precisely what happens—nothing, at least for one instruction cycle. In addition to other tasks, it is useful as a rather short time delay.

The next area we want to cover is the commands for arithmetic operations—that is, addition and subtraction. There are only four of these, but they will prove very convenient.

ADDWF, f,d: Like all of these instructions, this one is very straightforward. It simply takes the W register, or more appropriately what is in the W register, and adds it to a specified register. The result is then placed in the W or f register. Simple, eh?

ADDLW, k: This nifty command adds the a "literal" to the W register, and puts the outcome in W. That's all there is; there ain't no more!

SUBWF, f,d: Here, "SUB" refers to subtracting the W register contents from a particular register by "2's compliments". The result goes back into the W register.

Here are the commands for arithmetic operations— that is, addition and subtraction.

SUBLW, k: The name of this instruction is a little misleading, as it implies that you subtract the literal (L) from the working register (W). In reality, however, it is the other way around. W is subtracted from the literal. I stress this because misinterpreting this instruction is a very common programming mistake whose discovery and correction can result in a lot of wasted time. As with so may of these commands, the result is placed in the W register.

Our last, but not least, subject is the instructions involving digital logic (you know, AND, OR, NOT, etc.). These six commands are useful in manipulating data stored in the W register.

ANDLW, k: Here, the contents of the W register is "ANDed" with the literal (k), or "mask", and the result goes back into W.

ANDWF, f,d: As with the above command, the W register is "ANDed", but this time with the contents of a particular register. Again, the outcome goes to the W register.

XORLW, k: This one "exclusive OR's" ("XOR's") the W contents with the literal (k) and puts it into W.

XORWF, f,d: Like the ANDWF, this one "XOR's" the W register with the contents of a particular register. Results in W.

IORLW, k: Here, the W register is "Ored" with the literal (k). The W register then receives the outcome.

IORWF, f,d: And, this one "OR's" the W register with a particular register, with the result going back into W.

Hopefully, that gives you a better understanding of the PICMICRO® microcontroller instruction set. These are very simple commands that control the entire Microchip line.

However, you may be saying, "There are 37 instructions here, not 35." And, you are right! Thirty-five is the official number of commands, but most programmers are still using two "obsolete" instructions, TRIS and OPTION. This tends to make things easier, especially with the PIC16C54 chips. Believe me, though, this is a far cry from the instruction sets of some of the older microprocessors and microcomputers.

That's not to say those chips are not good devices. Many of them are excellent ICs, but their instruction set could have as many as 85 to 100 or more commands, and that made for a whole lot more work when using them.

It has been pointed out that this small instruction set, together with the 14-bit command size, has led to a number of compromises. While that may be true, the PICMICRO® MCUs have emerged as one of the best microcontrollers to hit the market in a long time. One very simple solution to the "instruction size" is to move everything through the working (W) register. That will remedy many a problem.

MORE DETAILS

So, now that you've mastered the instruction set … what? You have, haven't you? Well, anyway, now that you have a better comprehension of the instruction set, let me touch on some finer points that will be helpful when you start writing programs. You probably do have a good idea of what the various code variables are, but just in case, let me define them. Besides, this is good practice for me!

Let me touch on some finer points that will be helpful when you start writing programs.

"W": This stands for the "working" register.

"f": This is for the "file" register.

"d": Means "destination" register, and can be one of two places. If this is a "0", then the destination is the W register. If it is a "1", it goes to the file (f) register.

"k": Here, the reference is to a "literal instruction", or just plane "literal" (constant). If this is seen with respects to a "goto" or "call" instruction, however, then the "k" means an address label.

"b": This one refers to a "bit-oriented" instruction, or "bit designator".

That should explain what the letters mean in the various commands. For example, the "MOVF, f,d" instruction indicates that a chosen register's contents will be moved to a destination (d) of the "W" register or the "f" register. Likewise, "CALL, k" specifies calling a subroutine to a designated starting address, or literal (k).

As I go through this section, some of the discussion will refer to the PIC16F84 microcontroller, as this is one you will probably use often. Many of the devices in the Microchip line bear similar, if not identical, characteristics of the PIC16F84, however, so this discussion will apply to those chips as well.

To start with, just a reminder to be careful when typing programs into the text editor. If you're like me, mistakes (typos) are a common occurrence. They say it has something to do with my "misspent youth," but I don't know. Be that as it may, the program will not work if the code is not typed in correctly. (Another example of my overstating the obvious.)

However, simple typing mistakes are all too easy to make and very difficult, as well as time consuming, to find. For example, it is all too easy to get instructions backward, such as mistaking "loading W with the register" for "loading the register with W". They are *not* synonymous! Hence, be careful.

FLAGS

Okay, let me cover some of the highlights of programming the PIC16F84. There are others, but these I find to be most important. First, the microprocessor has three "flags", of which two are significant to us. They are the "zero" (Z) flag and the "carry" © flag. A flag is a one-bit register that provides information to the program.

For example, a flag will be "set" to indicate a certain action has taken place. The action is usually the result of a program instruction, and the program can "test" the flag to determine if it has been "set" or "cleared".

Setting a flag means giving the one-bit register a high, or "1" value, while clearing it means the opposite, a "0", or low value. The flag status can reflect one or more instructions, and in many cases, the direction the program will take depends largely on that status.

For the PIC16F84, flags are found in the "status word register" (f3). Thus, flags are very important to the operation of any program. As you work with the PICMICRO® microcontrollers, you will soon begin to comprehend the significance of the information the flags provide.

The PIC16F84 microprocessor has three "flags", of which two are significant to us. A flag is a one-bit register that provides information to the program.

LOOPS

Loops come in a variety of sizes, shapes, and flavors, so I will cover some of the more useful varieties.

This one we have talked about before, but I want to go into more detail concerning them. They also play a pivotal role in many programs, and do make programming sufficiently easier. Loops come in a variety of sizes, shapes, and flavors, so I will cover some of the more useful varieties.

The "endless" loop, as its name implies, is one that continues on and on "endlessly". Normally, the only way to stop this one is to shut down the power source. Since the PIC16F84 does not have a "halt" instruction, an endless loop can be used instead. To do this, use the "goto" instruction to take the program to a "circle" command, and the program will "circle" in an endless loop.

Another loop routine is one that loops until a certain action takes place. This "loop until" scenario allows for time delays and/or detection of an external activity by looping until said action occurs. Once the loop is interrupted, the program will then move on to the next command. For example, you might want a program to loop until a keystroke is detected, or a temperature sensor sends a reading to the microcontroller. In each case, the "loop until" routine will accomplish the required goal.

The last loop I want to discuss is a loop with a "counter". Here, the counter, which is usually a file register, is loaded with a specific number (literal). The loop routine then begins, and as it completes each cycle, the counter is advanced (incremented) by one. The program then tests the counter against the number of loop cycles to see if they match. If they do, or when the counter number equals the loop cycle number, the program either stops or moves on to the next instruction. The possibilities for this loop I think are obvious. No doubt, you will employ this type of loop on many occasions. It has also been pointed out to me that this loop demonstrates the ability of a microcontroller to make decisions—that is, being able to compare two numbers and act when they are identical (an important concept in computer science).

DELAYS

This subject closely parallels the loop; if fact, they are often called "time delay loops". Here, the program is stalled for a period of time as a result of a subroutine known as a "delay." As was illustrated in the last section, loops make great delays—hence, the name "time delay loop." Delays, of course, are employed whenever the program needs to "slow down" for a time.

An example is the LED flasher project in Chapter 4. Here, both a delay routine and a loop routine are called upon to provide a ½ second pause in the status of porta bit "0" (RA0). The delays (actually there are two of them) either hold the line high for ½ second or low for ½ second. The result being that the LED, which is connected to the port, turns "on" and "off," or flashes, at a 1-second rate.

Delays of less duration are both common and pragmatic in any program that needs the microcontroller to slow down. Such situations may involve the necessity of external data or an internal operation to "catch up" with the PICMICRO® microcontroller. Or, it may simply be a condition where the program—and microcontroller—has to wait for something to occur. Whatever the case, the delay will handle the task quite well.

PORTS

These are important! That is, if you want the microcontroller to do something useful. The ports are the lifelines to the outside world, and thus, are the PICMICRO® MCU's way of communicating with peripheral devices. Those devices can be any and everything from stepper motors to displays and keyboards to various sensing units. But, what is significant here is that the microcontroller would be nothing more than a long black rectangle with metal pins if it were not for the ports.

Looking at the PIC16F84, there are 13 port lines available, and they are arranged in two groups. The "A" port is five lines (RA0 to RA4) and the "B" port is eight lines (RB0 to RB7). They can be configured as input or outputs, and each line can be set individually. So, in addition to being essential, they are very versatile.

Input/output operation is determined by the TRIS instruction, with a "0" making a port line an output, and a "1" making said line an input. As you might remember, the W register's content will end up in the TRIS register when the TRIS command is executed, and the status of the lines (input or output) will be determined by TRIS.

The process of transferring information on the ports is called "reading" or "writing," depending on whether the port is in the input or output mode. A port is "read" at the beginning of the instruction cycle, but "written to" at the end of the cycle. If you plan to use a write instruction immediately after a read instruction, it is recommended that you separate the two with a NOP (nothing) command.

Recall that the NOP instruction results in the PICMICRO® microcontroller doing "nothing" for one instruction cycle, and that gives the system time to settle down before the write command is executed.

Hopefully, that gives you a good idea of how the ports function. Needless to say (again), they are critical to proper operation of any circuit using a PICMICRO® microcontroller or, for that matter, any type of micropro-cessor or microcontroller.

SUBROUTINES

By definition, any component of a program that is not part of the main routine is a subroutine. These are the small, inclusive routines that handle such duties as loops, time delays, lookup tables, and many others. And sub-routines allow the programmer to incorporate all kinds of neat stuff in his or her program. For the most part, it would be very difficult to write a program of any capabil-ity without these little gems.

Again using Chapter 4's LED flasher as an example, both the delays and the loops are subroutines to the main routine. Without those routines, the program would not function properly, the LED would not flash, and we might be the target of "the curse." So, the significance of the subroutines can easily be recognized. Usually, subroutines are accessed with a "goto" or "call"

Subroutines are the small, inclusive routines that handle such duties as loops, time delays, lookup tables, and many others.

instruction. This action takes the program to the subroutine, and it does its job. But a "return" instruction— and this is important—must be included as the last line of the subroutine. The return will take the code back to the main program so everything can continue.

On a last note regarding subroutines, they are extremely handy code, and it is a good idea to record them for future use. This can be done on the computer's hard drive or on a floppy disk. Since a subroutine for one program might very well be the ticket for another routine, you will save yourself a lot of time by having them available and ready to be copied into the new program.

LOOKUP TABLES

Also called "read tables," lookup tables allow you to place information into them, and then refer to that data anytime during the main program execution.

This is an approach that will not only save memory space, but will definitely simplify your code. I think so much of these tables that I have dedicated Chapter 6 to them. Also called "read tables," they allow you to place information into them, and then refer to that data anytime during the main program execution. A simple subroutine is used to access the table.

Let's say you want to build a device that will decode the Dual Tone, Multi-Frequency (DTMF) tone pairs found on your telephone line. On the surface, this might seem like a fairly complex project, but with the aid of a lookup table, it becomes quite manageable. The hardware will consist of a telephone line interface, a DTMF decoder chip, the PICMICRO® microcontroller, and a liquid crystal display (LCD) to annunciate the results.

The decoder chip, say a Motorola MC145436P, will receive the DTMF tones and then produce a 4-bit binary code representative of each separate pair of tones. Now the PICMICRO® MCU comes into play. The microcontroller will take that 4-bit nibble and convert it to American Standard Code for Information Interchange (ASCII) characters that the display understands. That, in itself, is not too bad.

But, how do we go about the binary code to ASCII conversion? The answer is simple. Each tone pair is going to be represented by the same binary code each time it is detected. So, all you need to do is set up a lookup table that recognizes a certain binary code as, let's say, the number "2", or "5", or whatever digit the code does designate. The main program will then access the table when necessary, and the display will receive the proper ASCII code for the numbers as they are detected.

There you have one example of how to use the "lookup" or "read" tables. I am certain you can conger up a number of other prime candidates. Along this same line, Morse code characters could be detected and decoded. If you are a HAM radio operator, that project could be very useful!

THE RETWL INSTRUCTION

This section is more of a "sidebar" to the above information on lookup tables than a subject of its own, but I think it will be useful. The "return from subroutine and load the W register with the literal" command (RETWL) will handle the task of accessing lookup tables.

In this scheme, after the RETWL instruction, the W register is loaded with the literal value.

What happens here is that when a RETWL is called, it accesses a part of the table contained in a subroutine. Following a series of "count" cycles, the counter data is loaded into the W register and the subroutine is called. The subroutine now transfers the W register information to the program counter, and the RETWL command holding the analogous code is called. The program returns to the main routine, with the W register containing the proper code. That, in turn, is sent to the LCD for display.

Hence, the RETWL instruction becomes an important part of using lookup tables. Like all of the PICMICRO® microcontroller software, it is simple to use and effectively does the job.

INTERRUPTS

Interrupts are a way of allowing the microcontroller to stop what it is doing (after it completes the current task) and proceed to another task that needs immediate attention. When that task is finished, the microcontroller is then free to return to what ever it was doing. This, of course, provides versatility in programming, as well as operation. Thus, interrupts are a standard feature with just about every microprocessor and microcontroller I can think of.

With the PIC16F84, an interrupt can be accomplished in one of three ways. First, an external influence can be applied to the RB0 port (pin 6) that serves double duty as the interrupt line. The interrupt will occur on the "rising" edge of the signal. Second, if the timer/counter TMRO experiences an overflow from 0xFF to 0x00. And a third possibility is changing the logic value (0 or 1) on portb lines 4, 5, 6, and 7. But, these have to have "input" status to be used as interrupt lines. In each case, an interrupt will be called (actually, there is a forth way that involves EEPROM writes, but it isn't applicable for our purposes).

Normally, interrupts are used when an external source, such as a sensor, needs to talk to the microcontroller, or when the time it takes for a flag to "set" is going to be excessive. Regarding the second situation, the timer can be used to generate the interrupt. At the end of the timer count, the associated flag is set, and the next instruction is placed on the stack. Referred to as a "goto the interrupt routine", this action will cause the PIC16F84 to jump to its 0x004 address and run the instruction present at that address.

Following execution of the interrupt routine, the "return from interrupt" (RETFIE) instruction will retrieve the original instruction from the stack, enable the interrupts, and the program can resume. So, you will want to have whatever instruction or routine you want to run in place and at address 0x004 before the interrupt is called. What becomes clear here is that proper interrupt operation depends on the "enable" and "flag" bits. To elaborate, with the "enable" bit, an interrupt may occur if the flag bit is "set". For the "flag" bit, the possible source of an interrupt has already occurred.

Incidentally, there are two types of interrupts—the "specific" and the "global." A specific interrupt is one in

Normally, interrupts are used when an external source, such as a sensor, needs to talk to the microcontroller, or when the time it takes for a flag to "set" is going to be excessive.

which a distinct, or "specific," source is the culprit, while a global interrupt refers to any interrupt experienced regardless of the source. The global variety is usually encountered more often than the specific.

THE PRESCALER

The prescaler is an 8-bit counter that can divide the clock input by one of eight binary values.

A nice feature of many of the Microchip microcontrollers is the "prescaler." This is an 8-bit counter that can divide the clock input by one of eight binary values—that is, the clock can be divided by 1, 2, 4, 8, 16, 32, 64, 128, or 256, thus reducing the clock frequency. I know, that is nine values, but the first, "1", actually bypasses the prescaler. This, in turn, assigns the prescaler to the "watchdog timer," and that is done to allow direct clock input of the timer/counter.

By the way, the watchdog timer is independent of the internal clock, and its main purpose is to keep a "leash" on the PICMICRO® MCU. That is, it helps prevent the microcontroller from getting away from you and not being able to recover.

Now, you ask, why would you want to divide the clock input by these various values? That's a very intelligent question! Give yourself a pat on the back! But, there is an equally respectable answer. (I hear you! You figured there was!) Actually, there are several respectable answers, but let's direct our attention to what happens when the clock input, to the timer/counter, is divided.

By dividing the input, you effectively slow the timer/counter down. The result here is that the resolution, in terms of frequency, is expanded proportionally to the division value selected. In other words, the PICMICRO® MCU can now handle signals with higher frequencies. That should ring some bells!

If your are familiar with the old "dedicated" prescaler chips of days gone by, then you will probably remember one of their primary uses. That was in "frequency counters" to reduce the frequency of the test signal. Since the counter ICs of that era had a limited operating frequency (usually 10 megahertz or below), if the frequency being measured was above that limit, it was necessary to reduce it. If you didn't decrease it, the dedicated counter could not give an accurate reading.

So, the prescalers were used to do just that. They had a frequency resolution of hundreds of megahertz (sometimes even in the gigahertz range) and could be cascaded, if necessary, to knock those higher frequencies down to where the counter could manage them. That meant a chip such as the Intersil/Harris ICM7216 dedicated frequency counter, with a 10-megahertz limit, could be used to measure frequencies in the hundreds, even thousands, of megahertz.

"All right," you say, "that's a nice history of the old frequency counters, but what does it mean in this context"? I was hoping you would ask that question, as that is what the "built-in" prescaler can do for the PIC16F84 and others. Over the past few years, I have encountered a number of articles about PICMICRO® MCU frequency counters, and many of the units could reach into the lower gigahertz range. Think about it—all that coming from a microcontroller with a base frequency of 4 megahertz! And that's just one example of how the prescaler can help you design pragmatic projects. Anytime you have a situation where the clock input needs to be reduced, the prescaler is your boy! Hence, keep "him" in mind for future endeavors.

ADDRESS MODES

This section is a quick look at the four "address modes" that are available for PICMICRO® microcontrollers. These are fairly straightforward, but they do need some attention. Let me start at the beginning, or with the "immediate" mode. "Immediate Addressing" refers to a literal instruction that has the data contained within that instruction. This will be loaded into the working register (W) and takes the data from the program memory—very basic.

The second mode, or "Direct Addressing," is just as basic, or maybe even more so, in that it specifies the location (address) after the instruction. An example would be "MOVWF, temp". Here, the contents of W are moved to the temporary (temp) register.

Third comes "Indirect Addressing." In this case, the "indirect address register," or first file register, is used. That has an address of f0, 0x00. However, that address doesn't exist, so the data actually goes to f4, 0x04. Hence, in this scheme, the 0x00 address will be replaced by the 0x04 position and used as a file register. This is a little complex, but it does make a good address pointer, and is useful when sending data to LCD displays.

Last is "Relative Addressing." This type of addressing results in a jump from the current address to a new address "relative" to the original address. For the PIC16F84, this encompasses altering the program counter contents.

COMPARISONS

Like loops, comparisons demonstrate the micro-controller's ability to make a decision. That is, the process of comparing two values and taking action depending on the result. In reality, that action may be doing nothing at all, but the point here is that the capability to do something is present, and that represents a very important concept displayed by microcontrollers.

Normally, the W register is compared with a literal value. This is accomplished by subtracting the contents of W from the literal. Whatever the result might be (the two are equal or not equal), the program can then branch to another instruction, or continue from where it left off. Either way, the comparison provides a useful service by keeping track of certain conditions and, when necessary, taking action based on those conditions.

In practical applications, comparisons are employed within the logic operation. They are used to determine such circumstances as "is W less than or equal to the literal, is W equal to the literal or is W greater than or equal to the literal", to name a few. These, of course, give the program a status report on where the literal and the W register stand.

That should help you further understand the intricacies of the PICMICRO® microcontrollers, especially that ol'

Comparisons demonstrate the microcontroller's ability to make a decision. That is, the process of comparing two values and taking action depending on the result.

When I say "structure," I'm referring to the arrangement of the text segments in a fashion that the MPLAB® assembler can recognize and know what to do with.

standby, the PIC16F84. Once you get to working with these devices, all this will be of greater importance—as well as make more sense. Trust me!

ASSEMBLER TEXT STRUCTURE

Now that you have reviewed the software and know what all the instructions do, it is time to use the code. So, let me touch on how this is done. Like so much involving the PICMICRO® microcontrollers, it isn't hard. It just requires the program text to have a certain structure. When I say "structure," I'm referring to the arrangement of the text segments in a fashion that the MPLAB® assembler can recognize and know what to do with.

I would imagine that since you are reading this book, you have an interest in PICMICRO® microcontrollers (you can see how brilliantly deductive my mind really is) and have probably seen some of the code by now. If you have looked over PICMICRO® MCU programs from different sources, you have also probably noticed a good deal of variance in the way the text looks.

Don't be concerned by this, as it merely reflects the style of whoever wrote the code. The important aspect, at least to the assembler, is that certain information be placed where it belongs. As long as that is right, you can add all the "window dressing" you want, and it won't bother the assembler one bit (excuse the pun).

Basically, what this involves is three columns of data. All the important stuff needs to go in one of those three columns, but what goes where does vary a little according to what type of instruction you are writing (i.e., header, program, equate, etc.). As I said, this is not hard to grasp; it just takes a little getting used to.

The order that the various types of instructions are placed in can be varied some, but an "absolute" is that each program must start with the "processor designator" and end with an "end" instruction. The other categories can be shuffled around, if desired, but personally I like to keep everything in the following order: Header, Equates, Origin, Program, End. By always using that placement, things stay nice and simple. If you hadn't already noticed, I'm a firm believer in simple.

Let's go through each instruction type and see what goes in each column. For Headers, the first column is left empty, the second column holds the "assembler directive," and the last column is the "processor description." So, this would look something like:

```
List        p=16f84
Radix       hex
```

Visualize that the page is broken into three equal columns. The first is empty, the directive is next, and the processor type is last. Next, the Equate command. Here, the first column is the "label," the second holds the "equate," and the third holds the "hex address or numerical value." It will look like this:

```
Porta       equ         0x04
```

Now, let's consider the Origin command. Column one is not used, two is the origin statement, and three is the hex address. Here's how it might look:

Org   0x0ff

Okay, on to the heart of the code, the program. The first column will be the "label" (if one is used), the second holds the "instructions," and the third column can be either a "literal" or a "label." Check this out:

Go   bsf   0x006

And the last, but essential, instruction will be "End." As I stated earlier, this command has to appear in all PICMICRO® MCU programs, and it will look like this:

End

All right, what is each of these telling us, as well as the assembler? The first line indicates the program is written for a PIC16F84 microcontroller. It further indicates that the format will be "hexadecimal," or "hex." Remember, this always has to be the first information the assembler receives.

Line two tells us that porta is assigned to the address "0x04", and line three verifies where the program starts— that is, at what address it starts. Bear in mind, Origin commands can be used in three different ways. As in this example, it can identify the starting address, but they also can be used to indicate the start of an interrupt or the reset vector for the PIC16F84.

Line four is, of course, the "meat" of the code. I only showed one command in the example, but this area will usually contain dozens of lines, sometimes hundreds. It is here the actual operational instructions are located.

The line shown tells the assembler to go to the selected bit at 0x006 and set it to "1".

Last, we have the very simple but important End statement. This merely tells us, and the assembler, the program is "ending." You know, "Thaaaaaaat's all, folks!"

Now, it is only this type of information, in its proper location, the assembler is interested in. So, why do you see so much other stuff in the average PICMICRO® MCU program? The rest of the text is purely for the benefit of you and I. That is, it labels the program, explains the lines of code, and/or indicates miscellaneous information about the program. Anytime you see the semicolon (;) at the beginning of anything within a PICMICRO® MCU program, that information will not be considered as part of the program by the assembler. In other words, it is ignored! Hence, the semicolon provides the programmer with a means of adding description, explanations, and other info to the text version of the program. Naturally, this helps us experimenters better comprehend the operation of said program. A good deal, don't you think?

Now, the above examples, if put together, would be a very short program that did nothing at all. It would look like this:

```
            List    p=16f84
            Radix   hex
Porta       equ     0x04
            Org     0x0ff
Go          bsf     0x006
            End
```

Anytime you see the semicolon (;) at the beginning of anything within a PICMICRO® program, that information will not be considered as part of the program by the assembler.

Trust me, this is useless! But, it does provide a simple (there's that word again) example of what each segment of the overall code is and does. What you will see, however, and probably already have seen, looks something like this:

```
;======PICTEST.ASM================DATE====
list        p=16f84
radix       hex
;------------------------------------------------------
porta       equ         0x04
;------------------------------------------------------
            org         0x000
start       movlw       0x00        ;load w with 0x00
            bsf         0           ;make RA0 high
            movlw       0x09        ;load w with 0x09
            movwf       porta       ; load ports with w
circle      goto        circle      ;finished
            end
;==========================================
```

Here, the program sets the porta lines as outputs, then sets the porta 0 line (RA0) as high, and finally puts the whole thing into a loop. A very simple routine, but it does serve to illustrate what a typical PICMICRO® MCU program might look like. Since the assembler will not recognize anything that starts with a semicolon, here is what it will see:

```
            List        p=16f84
            Radix       hex
Porta       equ         0x04
            Org         0x0ff
Start       movlw       0x00
```

```
              Bsf        0
              Movlw      0x09
              Movwf      porta
Circle        goto       circle
              End
```

Columnized code with everything in the proper place. Nice and neat! And that is the way all your programs will go together. Whether it is a simple routine like this one, or hundreds of lines long, the configuration is always the same. Follow the guidelines previously set forth, and formatting will not be a problem. Also, using programs that other people have written won't throw you. Their "style" (text appearance) may vary from these examples, but you can be assured that the instructions will conform to the "three columns, everything in its proper place" scheme, or the program will not work.

FLOW CHARTS

I haven't said much about "flow charts" in this book. In fact, I don't think I've said anything about them, but they are a way to lay out your programming ideas before you start actually writing the code. I personally don't use them very often, but then, some folks say I'm weird (you ought to hear what I say about them).

Anyway, flow charts are really a matter of preference. My philosophy on this is, if they make you feel more comfortable, then by all means use them. For many programmers, they save a lot of time and are considered a normal, if not essential, part of writing code. But, in my opinion, they are not mandatory.

Flow charts are a way to lay out your programming ideas before you start actually writing the code.

Basically, a flow chart is a block diagram of the software. It will illustrate the instruction sequence for the program through a series of boxes, diamonds, circle, etc., and it may well be worth the effort to draw the flow chart. The boxes and other shapes are labeled according to which part of the program they represent, and lines and arrows show the "flow" of the code. Again, for many, this definitely helps sort out the initial steps needed to construct the program properly.

So, for your edification, I have included a flow chart for the light-emitting diode (LED) flasher in Chapter 4. It is depicted in *Figure 11.1* and, hopefully, it will serve as an adequate example of what a flow chart looks like and how they work.

USING THE MICROCHIP MPLAB® IDE

This bundle of software is available from the Microchip home office and distributors (free of charge), or it can be downloaded from the Microchip web site.

The folks at Microchip, in an effort to make using the PICMICRO® MCU line even easier, have created a bundle of software they call the MPLAB® IDE (see *Figure 11.2*). This is available from the Microchip home office and distributors (free of charge), or it can be downloaded from the Microchip web site (http://www.microchip.com). One of the really nice aspects of this is that Microchip is constantly updating the software with new data, or as new products come on the market. So, you will be able to stay current by simply visiting the web site from time to time.

The first step in getting MPLAB® IDE ready is to install it on your personal computer (PC). There is nothing difficult about this. Simply follow the instructions on the CD-ROM or on the web site. The programs will be put into a folder

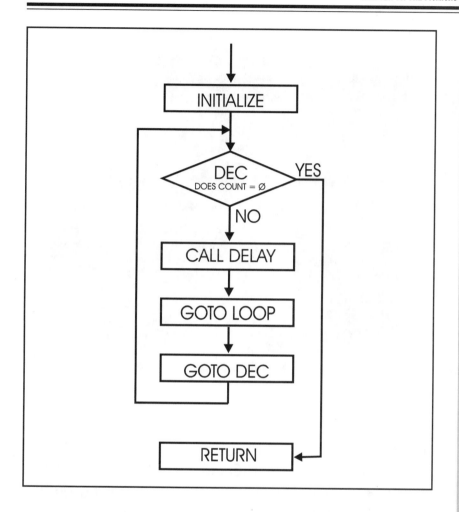

INITIALIZE

DEC
DOES COUNT = Ø

YES

NO

CALL DELAY

GOTO LOOP

GOTO DEC

RETURN

Figure 11.1.
Flow chart for the
LED flasher in
Chapter 4.

that can either be placed on the desktop or in the program directory. Incidentally, there is a MS-DOS version of MPLAB® IDE available, but unless your circumstances dictate it, I would highly recommend the Windows version instead. As of this writing, Microchip is no longer updating the DOS software.

Contained within MPLAB® IDE are three programs, or "development tools." These are a text editor, an assembler

Figure 11.2. The "MPLAB® IDE" and "Technical Library" CD-ROMs available from the home office of Microchip Technology Incorporated and its distributors, usually free of charge.

(MPASM™ assembler), and the control code for Microchip's dedicated "PICSTART® Plus" programmer. Having all this in one place, and in one folder, greatly simplifies the entire programming process.

The text editor is basically a word processor that produces clean ASCII format (no special control codes), and any such program will do. Since you have a great one built into MPLAB® IDE, however, why not use it? To get there, click on the "MPLAB" icon in the program directory, then on "MPLAB" in the dropdown list. Next, click on the "new project" box, and then on "new". This will bring up the text editor and a message asking if you want to create a new project. Click "yes". A box will come up asking for a directory and a name; put those in, and the cursor will begin to flash in the upper left corner of the text editor. You're ready! Type your program into the text area. When finished, click "file", and save the program to whatever place you want (hard drive, floppy disk, etc.). There are options on the toolbar that will allow you to edit and debug the code, and when you are done there, again save the software to the file you created.

Now, it is time to assemble the program. Go back to the dropdown box from the program directory and pick "MPASM for Windows". This will bring up a box that asks for all sorts of information. It will, of course, want the source file name, which you will type into the box. It will also ask for settings on such things as "radix", "warning level", "hex output", "macro expansion", and "generated files". In most cases, the default setting is going to be best; however, you can change the values if you choose.

At the bottom are two buttons, one of which you will use when you have finished with the settings. They are "Exit" and "Assemble". These are self-explanatory, and hopefully you will be ready to click on "assemble". By doing so, the MPASM™ assembler will assemble the code that you have stored in the file you have named—assemble it, that is, from what appeared in the text editor. This will result in a hexadecimal code file that can now be "burned" into the PICMICRO® microcontroller. There is another box at the bottom, and it asks if you want to "save settings on exit". I usually keep this one checked.

How you accomplish this last task will depend on your programmer. If you are using the Microchip PICSTART®

Plus programmer, the control software is built into MPLAB® IDE, and you can go back to the dropdown box and click on it. If you are using a programmer from another company or one that you have built yourself, then the control software for that particular device will now have to be accessed. It is best to load that code into the same folder you have MPLAB® IDE in. That way, it's always handy.

The first step, if you haven't already done this, is to connect the programmer to your PC. If the device connects to the PC's parallel or serial port, be sure to *shut down the computer* before trying to make the connection. Failure to do so might well result in damage or destruction to either the programmer, or the port, or both!

With your programming device hooked up and ready, run its software. Depending on the specific programmer, this will vary, but most of them will want information regarding such things as the file name, device being programmed, voltages used by the device, type of oscillator, is the watchdog timer "on" or "off", do you want code protection, and so forth. Make your choices in what is often referred to as the "status dialog box" and enter them. When you have completed the information-entry stage, move on to the programming stage. Here, the software will instruct the programmer to enter, or burn, the program into the PICMICRO® MCU's memory. It will usually run a test cycle or two to confirm the job has been done correctly, then inform you that it is finished. Normally, you will see a message such as, "Programming Completed Successfully".

Now, the PICMICRO® MCU is programmed and ready for action! Remove it from the programmer and move it over to the "prototype lab" we built in Chapter 2. By adding the various discrete components needed and making the appropriate connections to the peripheral devices (keyboard, display, etc.), the project will be ready to test— that is, test both the hardware design and the code.

CONCLUSION

Well, what do you think? Pretty neat, eh? This is about as good as it gets—at least, as of this writing. I have had my share of experiences over the years with many, many microprocessors, and I can honestly say the PICMICRO® microcontrollers are the easiest devices I have ever worked with. Not only is the software and hardware simple, but the extensive support efforts on the part of Microchip create an extremely functional environment within which to do the work.

All in all, I think your programming experience with these chips will be a very gratifying one. Unlike some of the predecessors, it will not be a constant struggle to achieve what you want. I have found that considerably fewer re-writes have to be done with the PICMICRO® microcontrollers. The code is that easy to use. Go get 'em, guys (of course, that is the generic "guys")!

I have had my share of experiences over the years with many, many microprocessors, and I can honestly say the PICMICRO® microcontrollers are the easiest devices I have ever worked with.

CHAPTER 12
CONCLUSION

This brings us to the end of our adventure with the Microchip PICMICRO® microcontrollers—at least, as far as this book is concerned, as I know it will not be the end of your interest in these devices. Nor will it be mine. These chips are so versatile and offer the hobbyist and engineer alike so much that they will continue to dominate the microcontroller market for years to come. But I know I have said all this before throughout this book, so let me not be redundant.

As you have undoubtedly gathered, the world of hobby electronics, as well as consumer electronics, is moving in this direction. So many projects we design and products for sale can be richly enhanced by the addition of "computer control." Hence, it's only natural that microcontrollers have caught on. And there is little doubt that any change in this philosophy is likely, at least in the near future. Thus, the more you can learn regarding microcontrol technology, the better off you will be in the long run. I have made every effort, over the years, to

Today, there is an abundance of information—in addition to this book—on the subject of microcontrollers, and especially the Microchip units.

keep up with progress in this area, and in this day and age of the Internet, such efforts are vastly easier. Also, the devices available now are downright simple to use as compared to their cousins of days gone by.

Today, there is an abundance of information—in addition to this book—on the subject of microcontrollers, and especially the Microchip units. Not only does Microchip provide an outstanding web site concerning their PICMICRO® MCUs, but the Web itself is full of sites dedicated to these marvelous little ICs. So, if you have questions regarding this subject, answers abound!

If you have Internet access, try going to a search engine such as "Dogpile.com" or "Search.com" and type in "PIC Microcontrollers" or "PICMICROs". I'm certain you'll be amazed at the number of hits you get. And just about all these sites will provide information, construction projects, or answers to your questions—sometimes all three. That, of course, makes working with the microcontrollers all the more fun. *Fun*—that's an important word, don't you think?

Additionally, as you gain confidence and experience, you will quickly be able to design just about any device you want. The magic of these chips lies in the ability to make, let's say, the PIC16F84 a dedicated integrated circuit. With that ability, your designs will become easier, more reliable, and *cheaper*! (You'll note that I keep my priorities straight!) On a final note of praise, I know I was somewhat intimidated by the early microprocessors and microcontrollers. It took me years of playing around with those devices to really get my "sea legs," so to speak, and that was due largely to the complex instruction sets. Writing code has never been one of my strong suits, but with the simple 37-command PICMICRO® MCU instruction set, writing code is now a delight. Here is something I can really understand and get a hold of! I love it!

Okay, what are you waiting for? With the knowledge you now have, you, too, can be designing PICMICRO® MCU projects. Hey, if I can do it, just about anybody can! It really doesn't take a deep or complete understanding of the PICMICRO® MCUs themselves. Just learn the code instructions and how to use them, and you will be on your way to a rewarding new facet of electronics.

I guess I had better quit, as I'm starting to sound like one of those "high-pressure" brochures trying to sell you something. Actually, that's appropriate, because I truly am trying to sell you on the PICMICRO® microcontrollers. Mark my words; you'll thank me for it!

Thanks for joining me, and have fun!

With the knowledge you now have, you, too, can be designing PICMICRO® projects. Hey, if I can do it, just about anybody can!

SOURCE LIST

This is a list of some of the possible sources for Microchip PICMICRO® microcontrollers and the development tools. Many of these companies I have dealt with in the past, thus I can vouch for them. The ones I have not had dealings with were advertised in reliable publications, so I trust them.

I have provided every means of contacting these companies that they advertise. Naturally, this is not every source available, so keep a lookout for new ones.

AdvancedTransdAtA
14330 Midway #128
Dallas, TX 75228
Phone: (972) 980-2667
Fax: (972) 980-2937
Web: www.adv-transdata.com
Programmers and Emulators

Many of these companies I have dealt with in the past, thus I can vouch for them.

Alltronics
2300 Zanker Road
San Jose, CA 95131
Phone: (408) 943-9773
Fax: (408) 943-9776
Web: www.alltronics.com
Parts, PICMICRO® MCUs, and Programmers

B.G. Micro
P.O. Box 280298
Dallas, TX 75228
Orders: (800) 276-2206
Tech Support: (972) 271-9834
Fax: (972) 271-2462
Web: www.bgmicro.com
Parts and Peripheral Devices

Circuit Specialists, Inc.
P.O. Box 3047

] GPS dev. kit

Scottsdale, AZ 85271-3047
Phone: (800) 528-1417
Web: www.cir.com
Parts and Programmers

DC Electronics
P.O. Box 3203
Scottsdale, AZ 85271-3203
Order: (800) 467-7736, or (800) 432-0070
Info: (602) 945-7736
FAX: (602) 994-1707
Parts and Programmers

∗ **Digi-Key Corporation**
701 Brooks Avenue
Thief River Falls, MN 56701-0677
Phone: (800) 344-4539
Web: www.digikey.com
Parts, Programmers, and PICMICRO® MCUs

∗ **Electronix Express**
365 Blair Road
Avenel, NJ 07001
Order: (800) 972-2225
Info: (732) 381-8020
FAX: (732) 381-1572
Web: www.elexp.com
Parts and Programmers

ITU Technologies
3704 Cheviot Avenue, Suite 3
Cincinnati, OH 45211
Phone: (513) 661-7523
Web: www.itutech.com
Programmers and Software

∗ **Jameco Electronics**
1355 Shoreway Road
Belmont, CA 94002-4100
Phone: (800) 831-4242
Cust. Service: (800) 536-4316
Tech Support: (800) 455-6119
Web: www.jameco.com
Parts, Programmers, and PICMICRO® MCUs

✳ **JDR Microdevices**
1850 South 10th Street
San Jose, CA 95112-4108
Order: (800) 538-5000
Fax: (800) 538-5005
Web: www.jdr.com
Parts, Programmers, and PICMICRO® MCUs

M2L Electronics
Durango, CO 81301
Phone: (970) 259-0555
Fax: (970) 259-0777
Web: www.m2l.com
Programmers

Marlin P. Jones and Associates, Inc.
P.O. Box 12685
Lake Park, FL 33403-0685
Order: (800) 652-6733
Fax: (800) 432-9937
Web: www.mpja.com
Parts and Programmers

✳ **Microchip Technology, Inc.**
2355 West Chandler Blvd.
Chandler, AZ 85224
Phone: (602) 786-7200
Fax: (602) 899-9210
Web: www.microchip.com
Complete PICMICRO® MCU Line, PICSTART® PLUS
Programmer, Software, and Tech Support

✳ **microEngineering Labs, Inc.**
P.O. Box 7532
Colorado Springs, CO 80933
Phone: (719) 520-5323
Fax: (719) 520-1867
Web: www.melabs.com
Programmers, Prototyping Stuff, and Software

Worldwyde.Com
33523 Eight Mile Road, # A3-261
Livonia, MI 48152
Phone: (800) 773-6698
Web: www.worldwydw.com
Programmers

GLOSSARY

Address   In this context, a memory location where data is placed.

Amp   Named after the French physicist Andre Marie Ampere, this is the basic unit of currrent.

Architecture   In this context, the physical structure of an integrated circuit. With the PICMICRO® MCUs, this is "Reduced Instruction Set Computing,"or RISC.

ASCII   Standing for "American Standard Code for Information Interchange," this is a standard computer character set used for data transfer. The LCD displays usually receive ASCII information regarding what they are to display.

Assembler   A computer program that changes the "text" version of a program into language, usually hexadecimal code, that the microcontroller can understand.

.ASM   The suffix that each assembler file will receive. Example :TEST.ASM.

Bank   No, not the First National, this is a group of file registers. Most PICMICRO® MCUs have two of them.

These important terms should aid your understanding of the topics in this book.

BCD   Standing for "Binary Coded Decimal," this is the language of computers based on multiples of 2. Example: 1, 2, 4, 8, 16, etc.

Binary   Refers to multiples of 2. Depicted as a string of 0's and 1's. Example: the number 10 would be 0000 1010.

Bit   In this context, this is one piece of the hexadecimal file. A "nibble" is made up of 4 bits and a "word" is 8 bits or more.

Breadboard   This term refers to a solderless method of wiring circuits for prototyping. The board has inter-connected sockets that the component leads are plugged into.

Byte   In this context, this is a "word" or 8 or more bits.

Capacitor   An electronic component that temporarily stores an electrical charge within a circuit. However, this is not synonymous with "battery."

Ceramic Resonator   An electronic component that consists of a quartz crystal and two small capacitors. These are used for clock timing with the PICMICRO® MCUs.

Chip   This is slang for "integrated circuit" and refers to the silicon matrix inside the IC.

Clock   In this context, this is the device that provides and maintains the frequency of the microcontroller cir-cuit. A quartz crystal or ceramic resonator is often used to control the clock, but a resistor/capacitor (R/C) tank can also be employed.

Code   This is another term for the software that controls the PICMICRO® MCUs.

Coil   Made up of turns of wire, this is an inductive component used in electronics often for filtering and/or frequency determination.

Command   Here, this refers to an instruction given the microcontroller.

Counter   A type of electronic circuit that counts pulses. These can also be used to divide an input signal.

Crystal   Made of quartz, these are used to define a frequency. A thin quartz "wafer" vibrates when exposed to electricity, and the vibration rate sets the frequency.

Data   This refers to just about any type of information put into, or received from, a circuit using a microcontroller.

DB connector   A standard style connector, with two offset rows, that is used in the computer industry. They come as DB9, DB25, DB36, etc., with the number referring to the total number of connections.

Delay   This refers to a subroutine that is called to delay the microcontrollers operation. Also known as a "delay loop."

Digital   In this context, digital refers to the nature of operation of devices associated with computers or microcontrollers.

Diode   An electronic device that changes an alternating current (AC) into a direct current (DC). These are commonly used in PICMICRO® MCU circuits.

DIP   Standing for "Dual In-Line Pin," this is a standard configuration for one style of IC and for some sockets. It consists of two rows of however many pins or sockets that are used in that particular arrangement.

DIP Switch   This is a small switch, usually used on a circuit board, that follows the DIP configuration and is compatible with that scheme.

Display   An electronic device, either liquid crystal (LCD) or light-emitting diode (LED), that is used to annunciate information.

EEPROM   Standing for "Electrically Erasable and Programmable Read Only Memory," this memory device can be electrically erased, while still in a circuit, and without the need of ultraviolet light. This is a nonvolatile type of memory, which means it doesn't lose the data when the power is disconnected. The PIC16F84 and other PICMICRO® MCUs make use of this type of memory.

End   This is an essential closing instruction for any PICMICRO® microcontroller program. As its name implies, it ends the program.

EPROM   Standing for "Electrically Programmable Read Only Memory," this type of memory has to be erased by shinning a high-intensity ultraviolet (UV) light through a small window in the chip onto the silicon wafer. This is a nonvolatile type of memory, which retains the data after power disconnect.

Equates   This is one of the basic segments of a PICMICRO® MCU program that usually defines an address, or assigns a label.

Erase   In this context, this is removing the information from a memory device. Random Access Memory (RAM) automatically performs this function when the power is removed, but EPROM and EEPROM memories require special erase procedures.

Farad   Named after Michael Faraday, this is the basic unit of capacitance.

File Register   Designated by the lowercase "F" (f), this is a place within a microcontroller for short-term data storage.

Flag   In this context, a flag is a single data bit, either high (1) or low (0), that indicates if something has or has not happened. These can be either set or cleared.

Files   This refers to a collection of instructions that tells a microcontroller to do something. These can be subroutines or complete programs.

Header   This usually refers to the top section of a program text that is used to identify the program and/or its operation. With the PICMICRO® MCUs, these lines begin with the semicolon to prevent them from interfering with the assembler.

Henry Named after Joseph Henry, this is the basic unit of induction.

Hertz Named in honor of German electrical pioneer Heinrich Hertz, this is the basic unit of frequency.

Hexadecimal Often referred to simply as "hex," this is a form of writing data that the assembler uses and is understandable by the microcontroller. Most of the PICMICRO® MCU files will be hex files.

Inductor Another name for a coil, this is a component that consists of turns of wire wrapped around a core that can be metallic, nonconductive, or just plain air.

Input This refers to anything that goes into the microcontroller, as opposed to leaving the microcontroller. The port lines are set up as either inputs or outputs (I/O), and they handle most of these duties.

Instruction In this context, an instruction is code that tells the microcontroller to do something. Programs are made up of these instructions.

Instruction Set Each microcontroller or microprocessor needs a basic group of special instructions that it understands to do its job. This group of special commands is known as the instruction set.

Integrated Circuit Also known as "IC," this is a semiconductor device that contains a number of other semiconductors such as diodes and transistors. These are etched into a silicon wafer and are often a complete circuit within themselves. Access to the circuit is supplied by metal pins that protrude from the case.

Internet   This is a network of connected computers that provides great access to information from the private sector, educational institutions, and local, state, and federal government organizations. Very handy!

Interrupt   In this context, this is a command or instruction that causes the microcontroller to stop what it is doing and direct its attention to another task. After the task is completed, the controller returns to what it was doing.

Key   Usually referring to a switch, these are used to input data to a microcontroller circuit.

Keypad   This is a collection of switches that is used to input data to a microcontroller circuit. These can be wired individually, in series, or as a "X/Y" matrix.

LCD   Meaning "liquid crystal display," this is a low-power form of a digital display. These use liquid crystals that twist and turn dark when electricity is applied, thus indicating the information.

LED   Meaning "light-emitting diode," these are special purpose diodes where the gallium/arsenide junction glows when electricity is passed through them. They are also seen as complete seven-segment displays that light up red, green, orange, or yellow.

Literal   Here, literal means literally the value. As part of the PICMICRO® microcontroller instruction set, they are used to place a specific value in a specific place.

Lookup Table   This is a special table that holds information that a PICMICRO® MCU program can refer to. This might be Morse code characters, DTMF telephone tone pairs, or any other information a program would want to frequently access. These save a lot of memory space, as the program doesn't have to keep repeating the data.

Loop   Here, the term loop is used to describe a sub-routine that just keeps going around and around in circles. These can be continuous, or stop after a certain parameter is reached.

Memory   In this context, memory is the ability of the microcontroller, through the use of several different devices, to store information it needs or will need. RAM, EPROM, and EEPROM are three common examples of memory storage devices.

Microchip Technology Inc.   This is the company responsible for developing the PICMICRO® microcontrollers and associated development tools.

Mode   In this context, mode refers to the operational status of an instruction or the microcontroller itself.

MPASM™ Assembler   This is the name given to the assembler program built into the MPLAB development tool package.

MPLAB® IDE   Microchip Technology's development tool package for working with the PICMICRO® microcontroller chips.

MS-DOS   An older operating system (OS) that is the basis of the windows operating system. Not used much anymore.

Multiplex   Here, multiplex refers to connecting elements of a circuit together to conserve input/output port lines.

For example, a multiplexed LED display will connect all "a", "b", "c", etc., segments of each individual display unit to a common line and access each unit with the common anode or common cathode line. This keeps the number of lines to the microcontroller to a minimum.

Nibble (sometimes "Nybble") This is a 4-bit segment of code, as opposed to an 8- or larger bit segment that is referred to as a "byte," or "word."

Ohm Named after German mathematician George Simon Ohm, this is the basic unit of resistance.

Option Register A special register which, in the PIC16F84, controls the watchdog timer, timer/counter, port b pullup resistors, and interrupt signal edge select. Hence, this is an important register.

Origin Abbreviated "ORG," this is one of the basic program segments that controls the program start address, reset vectors, and the start of interrupt routines.

Oscillator This is the electronics of the device that provides the microcontroller operating frequency. It can be controlled by a quartz crystal, ceramic resonator, resistor/capacitor network, or an external clock source.

Output As with "input," this can be the status of any of the PICMICRO® microcontroller port lines. In the output configuration, data is taken from the PICMICRO® MCU via the ports.

Package Don't laugh, but this is the type of case the microcontroller is housed in. I only bring it up because Microchip Technology offers a wide variety of said packages. These would include ceramic, plastic, PLCC, and so forth.

PCB   Standing for "Printed Circuit Board," this is a type of wiring in which the connection wires are actually traces of copper metal on a piece of nonconductive base material. Holes are drilled in the base material to run the component leads through, to be soldered on the trace side, or in the case of surface-mount technology (SMT), the traces and components are on the same side of the board, with the components being soldered directly to the traces.

Perf-Board   This is a type of wiring scheme in which the component leads are pushed through evenly spaced predrilled holes and connected with lengths of hookup wire.

Port   This is any line on a PICMICRO® microcontroller that is used to input or output data. The I/O status of these lines is set by the software.

Potentiometer   This is a variable resistor that is used to control such things as LCD display contrast and audio volume.

Prescaler   With the PIC16F84, this is a built-in device that divides the clock input by a binary value. That value can be 1, 2, 4, 8, 16, 32, 64, 128, or 256 and is determined by the software. Useful for a number of applications.

Program   This is the "bundle" of code that collectively controls the PICMICRO® microcontroller. The bundle consists of the various instructions, or commands, needed for the PICMICRO® MCU to do its job.

Programmer   The electronic device that places, or "burns," the control code into the memory of a PICMICRO® microcontroller.

Project   This is the design for which you are writing the code. It is also the term used by MPLAB to designate the file into which the software will go. Each time you start a new program, the assembler will ask for the project name.

RAM   Standing for "Random Access Memory," this is a volatile type of memory that is used to store data in a short term fashion. This memory will lose whatever is in it when power is removed.

Read   This is a process by which data within the microcontroller is retrieved to an external location. This is done through the chips ports.

Read Table   Another term for a "lookup table," this is a table that contains frequently used information.

Register   There are a number of these in most microcontrollers, and they are small memory spaces used to transfer or temporarily store data.

Regulator   In this context, a "regulator" is an electronic device (IC) that keeps the operational voltage constant no matter how much the input voltage varies. For microcontroller circuits, this is very important.

Reset   This is simply the process of resetting the microcontroller. It is done when a problem occurs or to clear all internal data.

Resistor   This is an electronic component that "resists" electrical flow. Used to maintain proper bias levels or as part of an oscillator tank circuit.

Routine   A term that describes instructive code—that is, code that commands the microcontroller to do something.

Semiconductor   One of a variety of electronic components, such as diodes and transistors. These are at the heart of microcontroller and get their name from their property of only conducting a certain amount of the total electricity available.

SIP   Standing for "Single In-Line Pin," this is a single row of sockets or conductors evenly spaced.

Sleep   This term refers to putting the microcontroller into a very low power mode. This is often done to conserve battery power.

Software   A collection of instructions that controls the microcontroller. With the PICMICRO® MCUs, this is placed in internal memory within the chip.

Source Code   Another name for software, this is the control code.

Stack   This is special purpose register within the microcontroller that handles the task of keeping an instruction until it is needed. Instructions come back off the stack in a certain order and are often the commands that were set aside due to an interrupt.

Status Register   A special purpose register that maintains the "status" of various flags and other data bits. This can be referred to by the program for operational information.

Subroutine   This is any block of software that is not part of the main routine or program. These are used to execute small programs within the main program by "calling" them up as needed.

Text Editor   This is a word processor used to write the text version of a PICMICRO® MCU program. One is built into MPLAB® IDE, but others can be used as long as they are ASCII in nature, without special control codes.

Time Delay Loop   Synonymous with "loop," this is a subroutine that continues on and on to delay the program.

Timer/Counter   This is a multifunction inclusion that has a number of nice features. Naturally, as the name implies, its main function is counting and timing, but it also has the ability to read and write data, use an internal or external clock source, and has an 8-bit programmable prescaler. To fully understand this device, you do need to read further, as there is just not enough space here to do it justice.

Transistor   This is a semiconductor device that is used for oscillating, amplifying, and switching. Basically, the potential on the "base" connection controls the degree of flow within the transistor.

Ultraviolet Light   A form of light, with a wavelength just above visible light, that is used to erase EPROM memories. This is done by the UV light supplying the energy necessary to replenish the areas of the memory that were depleted in programming. In this fashion, the entire memory area is returned to a uniform value.

Vector   This refers to a "standard" setting or location within the microcontroller's memory. For example, the "interrupt vector" is 0x04, and each time an interrupt occurs, the vector will start the new routine at that address.

Volt   Named after Italian physicist Count Alessandro Volta, this is the basic unit of electricity.

Wakeup   This refers to the process of returning the microcontroller from a "sleep" mode.

Watchdog Timer   The purpose of this timer is to keep the microcontroller from straying off into a condition where it cannot recover. In other words, this keeps the PICMICRO® MCU from locking up.

Word   This is another name for "byte," or eight or more bits grouped together.

Working Register   Also known as the "W" register, this is one of the hardest "working" registers in the microcontroller. Often, just about everything goes through the working register, and it is an essential part of placing data in memory. This is one important register!

Write   This is the process of placing data into the microcontrollers memory through the port lines. You will encounter the phrase "writing to memory" often as you work with microcontrollers.

INDEX

C

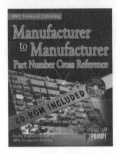

DSP Filters

John Lane & Ed Martinez

Digital filters and real-time processing of digital signals have traditionally been beyond the reach of most, due partially to hardware cost as well as complexity of design.

In recent years, however, low-cost digital signal processor (DSP) development boards have put the technology well in reach. This book, part of Sams Technical Publishing's "Electronic Cookbook series," will break the design-complexity barrier by means of simplified tutorials, step-by-step instructions, and a collection of audio projects. Design formulas are presented to build the digital equivalent of standard audio filters: lowpass, highpass, bandpass, and banstop.

Author John Lane is a staff scientist with the Image and Entertainment Division of Motorola, and he has been involved in software-algorithm development of IIR filters for audio, adaptive filtering, and music synthesis. Author Ed Martinez is a product manager with the DSP Tools Organization in the Motorola Core Technology Center in Melbourne, Fla., and has worked for Motorola for more than eight years.

Electronics Technology
336 pages • paperback • 7-3/8" x 9-1/4"
ISBN: 0-7906-1204-6 • Sams 61204
$39.95

Advancing to Microsoft® Office 10

John Breeden & Michael Cheek

This book provides valuable insight into the newest product from Microsoft®—Office 10. The replacement for Microsoft® Office, Office 10 is designed to take users into the 21st century.

Authors John Breeden and Michael Cheek provide tips and tricks for the experienced office user, helping you to get the maximum value out of this new software. Key features of *Advancing to Microsoft® Office 10* are: custom installation instructions; full coverage of Excel, Word, PowerPoint, Outlook, and more; complete explanations of all the new software features not found in previous versions of Office.

Breeden and Cheek are editors at *Government Computer News*, a trade publication owned by the *Washington Post*, and are experts in the fields of computer hardware and software.

Release of this book is entirely dependent on the release date of Office 10 from Microsoft®, which is subject to change at any time.

Business Technology
304 pages • paperback • 7-3/8" x 9-1/4"
ISBN: 0-7906-1233-X • Sams 61233
$39.95

To order today or locate your nearest Prompt® Publications distributor at 1-800-428-7267 or www.samswebsite.com

Prices subject to change.